Fundamentals of
Plastics Thermoforming

Synthesis Lectures on Materials Engineering

Fundamentals of Plastics Thermoforming
Peter W. Klein
2009

Fundamentals of Plastics Thermoforming

Peter W. Klein

ISBN: 978-3-031-01264-8   paperback
ISBN: 978-3-031-02392-7   ebook

DOI 10.1007/978-3-031-02392-7

A Publication in the Springer series
SYNTHESIS LECTURES ON MATERIALS ENGINEERING

Lecture #1

Series ISSN
Synthesis Lectures on Materials Engineering
ISSN pending.

Fundamentals of Plastics Thermoforming

Peter W. Klein
Ohio University

SYNTHESIS LECTURES ON MATERIALS ENGINEERING #1

ABSTRACT

The process of heating and reshaping plastics sheet and film materials has been in use since the beginning of the plastics industry. This process is known as thermoforming. Today this process is used for industrial products including signage, housings, and hot tubs. It also produces much of the packaging in use today including blister packs, egg cartons, and food storage containers. This process has many advantages over other methods of producing these products, but also it has some limitations. This book has a twofold purpose. It is designed to be used a text book for a course on thermoforming. It is also intended to be an application guide for professionals in the field of thermoforming including manufacturing, process and quality engineers and managers. This book is focused on process application rather than theory. It refers to real products and processes with the intent of understanding the real issues faced in this industry. In addition to materials and processes, part and tool design are covered. Quality control is critical to any operation and this is also covered in this text. Two areas of focus in today's industry include Lean operations and environmental issues. Both of these topics are also included.

KEYWORDS

thermoforming, plastic memory, morphology, Van der Waals' Forces, polystyrene, Acrylonitrile Butadiene Styrene (ABS), Polyvinyl Chloride (PVC), Poly Methyl Methacrylate (PMMA or Acrylic), High Density Polyethylene (HDPE), Low Density Polyethylene (LDPE), polypropylene, cellulosics, Polyethylene Terephthalate (PETE or PET), green plastics, heat deflection temperature, glass transition temperature, melt temperature, polymer set temperature, homopolymer, co-polymer, thermoforming window, mechanical forming, vacuum forming, pressure forming, drape forming, pneumatic preforming, billow forming, mechanical preforming, plug assist, twin sheet thermoforming, draw ratios, aerial draw ratio, linear draw ratio, height-to dimension ratio, corners and radii, draft angles, depth of draw, webbing, mold materials, mold geometry/shrinkage, venting, cavities, regrind, sheet orientation, variation, eliminating waste, 6S, process flow chart, process mapping, visual controls, poke yoke

Dedicated to my loving wife, children, and grandchildren.

Contents

SYNTHESIS LECTURES ON MATERIALS ENGINEERING iii

Contents ix

1 Introduction 1
   1.1 Introduction to Thermoforming 1
        1.1.1 Common Thermoformed Products   1
   1.2 Why use Thermoforming? 1
        1.2.1 Advantages of Thermoforming   1
        1.2.2 Limitations/Disadvantages of Thermoforming   4
   1.3 Plastic Memory 5

2 Plastics Materials 8
   2.1 Introduction 8
   2.2 Molecular Structure 9
   2.3 Morphology 10
   2.4 Van der Waals' Forces 11
        2.4.1 Discovering Van der Waals' Forces   11
   2.5 Most Used Materials 12
        2.5.1 Polystyrene (PS)   12
        2.5.2 Acrylonitrile Butadiene Styrene (ABS)   13
        2.5.3 Polyvinyl Chloride (PVC)   13
        2.5.4 Polymethyl Methacrylate (PMMA or Acrylic)   13
        2.5.5 High Density Polyethylene (HDPE)   13
        2.5.6 Low Density Polyethylene (LDPE)   14
        2.5.7 Polypropylene (PP)   14
        2.5.8 Cellulosics   14

2.5.9 Polyethylene Terephthalate (PET)   14

2.5.10 Green Plastics   14

2.6   Important Terminology 14

2.6.1 Heat Deflection Temperature (HDT)   14

2.6.2 Glass Transition Temperature (Tg)   15

2.6.3 Melt Temperature (Tm)   15

2.6.4 Polymer Set Temperature   15

2.6.5 Homopolymer   16

2.6.6 Thermoforming Window   16

3   Thermoforming Process Overview 17

3.1   The Thermoforming Process 17

3.2   Sheet Preparation 17

3.2.1 Drying   19

3.3   Loading 20

3.4   Heating 20

3.5   Forming 22

3.6   Cooling 23

3.7   Unloading 24

3.8   Trimming 25

4   The Forming Process 27

4.1   Forming Introduction 27

4.2   Mechanical Forming 27

4.3   Vacuum Forming 29

4.4   Pressure Forming 29

4.5   Combination Forming Processes 30

4.5.1 Drape Forming   30

4.5.2 Pneumatic Preforming   31

4.5.3 Mechanical Preforming   31

4.6   Twin Sheet Thermoforming 32

4.7   Laminating Thermoforming 32

5   Part Design 35

5.1   Design Questions 35

5.2   Wall Thickness Variation 36

5.2.1 Plug versus Cavity   36

5.3   Draw Ratios 41

5.3.1 Aerial Draw Ratio (ADR)   41

5.3.2 Linear Draw Ratio (LDR)   43

5.3.3 Height – To - Dimension Ratio   44

5.4   Material Selection 44

5.5   Part Geometry 45

5.5.1 Corners and Radii   45

5.5.2 Draft Angles   46

5.5.3 Depth of Draw   47

5.5.4 Webbing   48

5.5.5 Undercuts or Negative Draft   48

5.6   Part Application Issues 48

5.6.1 Useful Temperature Range   49

5.6.2 Strength Requirements   49

5.6.3 Stiffness   49

5.7   Quality Requirements 50

5.7.1 Cosmetics   50

5.7.2 Optics   50

5.7.3 Dimensional Tolerances   51

6   Mold / Tool Design 53

6.1   Purpose of the Mold or Tool 53

6.2   Mold Materials 53

6.3   Mold Geometry / Shrinkage 53

6.4   Venting 55

6.5   Temperature Control 57

6.6   Cavities 57

7   Quality Control Issues 61

7.1   Introduction 61

7.2   Material Source 61

7.2.1   Regrind   61

7.2.2   Sheet Orientation   62

7.2.3   Orientation Test   62

7.2.4   Moisture   64

7.2.5   Cosmetics   64

7.3   Process Quality 65

7.3.1   Machine Set-up   65

7.3.2   Variation   65

7.3.3   Tool Quality   66

7.3.4   Facility Quality   66

7.4   Quality Inspection 66

7.4.1   Quality Tools   67

8   Lean Operations 71

8.1   Introduction 71

8.2   Eliminating Waste 71

8.3   6S 72

8.4   Process Flow Chart 72

8.5   Process Mapping 72

8.6   Visual Controls 73

8.7   Set-Up Reduction 74

8.8   Poka Yoke 75

9   Environmental Issues 77

9.1   Introduction 77

9.2   Source Reduction 77

9.3   Recycling 77

9.4   Material Selection 78

9.5   Green Plastics 79

About the Author 83

CHAPTER 1

Introduction

1.1 INTRODUCTION TO THERMOFORMING

Thermoforming is an industrial process in which thermoplastic sheet (or film) is processed into a new shape using heat and pressure. This was one of the earliest processes to be used in the plastics industry beginning with the forming of cellulose nitrate sheet in the mid 1800's. The growth increased dramatically as new materials and applications were developed. For example, the need for aircraft canopies in World War II along with the development of polymethyl methacrylate (acrylic) created the perfect opportunity to advance thermoforming process technology. A growth rate of approximately 5% to 6% has been sustained for over forty five years.

Today this process is used to produce many products from small blister packs to display AAA size batteries to large skylights and aircraft interior panels. The market is often defined by the end use of the products being manufactured. "Industrial Products" include items with expected long life such as those used in the transportation and construction industries. "Disposable Products" (non-packaging) include items that have a short life expectancy but are not in the packaging side of the business. This market includes disposable plastic plates and drinking cups. "Packaging Products" is a huge, high volume, industry devoted to providing manufacturers with low cost packaging to display, protect, and/or extend the life of their products.

1.1.1 COMMON THERMOFORMED PRODUCTS
See the Figs. 1.1–1.7.

1.2 WHY USE THERMOFORMING?

The decision to select thermoforming as the process to manufacture a product has many dependencies. Some of these are covered in the Part Design Chapter 5. However, there are a number of general advantages and limitations/disadvantages that can be discussed.

1.2.1 ADVANTAGES OF THERMOFORMING

Low Equipment Costs: Thermoforming processing equipment is relatively low cost compared to other plastics processing equipment such as injection molding. The force required to form a sheet is typically less than 14 PSI for vacuum forming and less than 150 PSI for pressure forming. This compares with over 10,000 PSI for injection molding.

Low Tooling Costs: Due primarily to the low pressure used in the processes of thermoforming, tooling/molds can be made from a variety of materials. Although aluminum is most common for production molds, low volume molds can be produced from wood, tooling plaster such as Hy-

Industrial

Figure 1.1: Spa Liner

Figure 1.2: Aircraft Canopy

Figure 1.3: Backlit Signage

drocal B11, epoxy, composites and other materials that can withstand the pressures as well as the temperatures used in this process.

Economical to Produce at Low Volumes: Tooling costs are low, especially with prototype tooling; therefore, low volume products can be economically produced. The tooling costs must be amortized over the volume of parts produced: lower tooling costs equal lower product costs.

Timeline from Design to Prototype: Low cost tooling, especially wood, plaster and composites, are very quick to produce. A model maker may be able to produce a simple low volume tool in only a few hours so an actual part can be reviewed very quickly.

Disposables

Figure 1.4: Disposable Products

Figure 1.5: Disposable Foam Products

Packaging

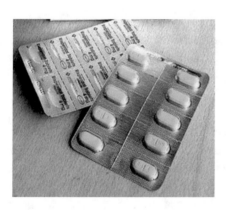

Figure 1.6: Blistering Packaging

Figure 1.7: Food Packaging

Large Surface to Thickness Ratios Common: Very thin parts, such as disposable drinking cups, can be produced which are too thin for processes such injection molding to produce. Also large products such as refrigerator liners (Fig. 1.1–1.8) and hot tub shells can be produced which may exceed the capabilities of common molding processes.

Wide Variety of Plastics Available: Almost all thermoplastics that can be created into a sheet can be thermoformed.

Figure 1.8: Refrigerator Liner   Figure 1.9: Multi-Layer Parts

Decorating and Printing: Common printing processes are often performed on the sheet before the forming process. Printing after forming is also economical and provides tremendous freedom for the product designers as well as a very low cost method to decorate the product.

Multi-Layer Parts: Extruded sheet can be produced with many layers of materials including different colors, grades, additives, etc. This can result in multi-layered thermoformed products. A common example is a disposable cup with a white interior and colored exterior. (Fig. 1.9)

1.2.2   LIMITATIONS/DISADVANTAGES OF THERMOFORMING

Non-Uniform Wall Thickness: This is the number one disadvantage of the thermoforming process. Since thermoforming is a "stretching" process, wall thickness of the product varies depending on the amount of stretching that must occur to create the desired geometry. There are many design rules as well as process variations to lessen the impact of "stretching." These will be covered in future chapters.

Sheet Cost: Thermoforming is considered a secondary process as the sheet must first be produced. The sheet manufacturing costs add to the raw (incoming) material costs for thermoformed products. Sheets are typically produced using extrusion, casting or calendaring processes.

Trimming: The sheet from which the part is to be thermoformed must be secured in a frame or clamp during the heating and forming process. After forming, the parts must be removed from the sheet using one of many trimming operations. Some of these are highly automated and very fast requiring expensive equipment such as robots and lasers. Others may require extensive labor to manually trim each part.

Trim Material Cost: "Trim" is the remnant left after the part is removed from the formed sheet. These skeletal remains may be as much as 80% of the original sheet material. This material is typically reclaimed using grinding and reprocessing methods which add more costs to the product.

Part Geometry Limited: Most of the thermoforming processes use a mold in which the sheet is stretched and cooled. This single-surface mold creates detail on only one surface of the part. Also undercuts and hollow objects, although possible, create added difficulties as well as costs.

Useful Temperature Range: Since thermoforming is a forming, not molding, process the product has "memory." The molecules are held in a state of stress and if the temperature reaches the heat distortion temperature (often similar to the lower forming temperature) the part will return to its sheet form.

1.3   PLASTIC MEMORY

As mentioned above, thermoformed parts have plastic memory. This is a result of the molecules being relocated without flowing as they do in processes like injection molding. The forming temperatures are below the temperature where molecules can flow, therefore the forming processes force the material into a new shape and freezes it in that shape when the heat is removed. The following activity can be performed on any thermoformed part:

1. Determine the material type. On packaging and disposables, this should be identified on the part using the recycle symbol, number and abbreviation for the polymer used. These are indicated in Table 1.1.

Table 1.1: Packaging Material Recycle Label	
#1 PETE	Polyethylene Terephthalate
#2 HDPE	High Density Polyethylene
#3 V	Polyvinyl Chloride
#4 LDPE	Low Density Polyethylene
#5 PP	Polypropylene
#6 PS	Polystyrene
#7 OTHER	None of the Above

2. Determine the lower forming temperature. If the material is identified as OTHER, further investigation is needed to identify the material so forming temperature can be determined.

   Forming temperatures (degrees F) for select materials are shown in Table 1.2.

3. Preheat an oven to the lower forming temperature.

   NOTE: This activity may not work well for many crystalline materials including polyethylene and polypropylene as the forming temperature and melt temperature may be very close.

Table 1.2: Forming Temperature for Select Materials			
Material	Lower Forming Temp	Normal Forming Temp	Upper Forming Temp
ABS	260	300	360
Acrylic	300	350	380
Polycarbonate	335	375	400
PETE	250	300	330
HDPE	260	295	360
LDPE	255	285	350
PP	290	320	330
PS	260	300	360
PVC	220	280	310

4. Place the thermoformed product to be tested on a non-stick surface and place it in the oven for 1 minute. (Fig. 1.10)

5. Check part status. If the part has flattened, remove from oven, if not, continue test for another minute. (Figs. 1.11 and 1.12)

Figure 1.10:   Figure 1.11:   Figure 1.12:

6. If part has not de-formed, increase temperature to normal forming temperature and repeat step 4 and 5 until part has flattened.

Results: Only thermoformed parts have the internal stress which is relieved by heating to its forming temperature. In addition to thermoforming, blow molding is also a stretching process and has similar issues with internal stress in parts produced.

Figs. 1.13 through 1.15 show another test example.

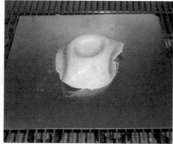

Figure 1.13:

Figure 1.14:

Figure 1.15:

CHAPTER 2

Plastics Materials

2.1 INTRODUCTION

The intent of this chapter is to provide a brief overview of plastics as an industrial material as it relates to thermoforming rather than a study of polymer science. Today's industrial plastics are synthetic, organic, hydrocarbons made primarily from petroleum. Their molecules are very long in relation to their diameter. The word polymer, used in this book synonymously with the word plastics, is derived from the Greek word poly, meaning many, and meros, meaning units or parts. The units or parts in the polymer are called monomers and are the initial molecule used in forming the polymer. For example, many ethylene monomers (at least 500) are joined together through a process called polymerization, to form polyethylene (in this case low density polyethylene). As the number of ethylene monomers increases, so does the density thus becoming medium density, high density and ultra high density polyethylene.

$n = 1$ = ethylene monomer – a gas

$n = 6$ = liquid

$n = 36$ = grease

$n = 100$ = wax

$n = 500$ = low density polyethylene plastic

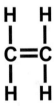

Figure 2.1: Ethane

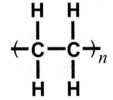

Figure 2.2: Ethylene Monomer

There are two primary categories of plastics, thermoplastics and thermosets. Thermoplastics are materials that can be heated, softened, reformed, and cooled to a solid state many times. This is often compared to candle wax. Thermosets use heat (internal or external) to cure the material and once cured, they cannot be softened and re-formed. A common, non-polymer, example is concrete which goes through a chemical reaction generating internal heat and curing the material. Concrete cannot be ground up and re-cast. About 80% to 90% of the plastics used are thermoplastics and these are the materials used in the thermoforming process.

C H A P T E R   2

Plastics Materials

2.1   INTRODUCTION

The intent of this chapter is to provide a brief overview of plastics as an industrial material as it relates to thermoforming rather than a study of polymer science. Today's industrial plastics are synthetic, organic, hydrocarbons made primarily from petroleum. Their molecules are very long in relation to their diameter. The word polymer, used in this book synonymously with the word plastics, is derived from the Greek word poly, meaning many, and meros, meaning units or parts. The units or parts in the polymer are called monomers and are the initial molecule used in forming the polymer. For example, many ethylene monomers (at least 500) are joined together through a process called polymerization, to form polyethylene (in this case low density polyethylene). As the number of ethylene monomers increases, so does the density thus becoming medium density, high density and ultra high density polyethylene.

n = 1 = ethylene monomer – a gas
n = 6 = liquid
n = 36 = grease
n = 100 = wax
n = 500 = low density polyethylene plastic

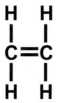

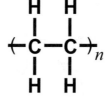

Figure 2.1: Ethane   Figure 2.2: Ethylene Monomer

There are two primary categories of plastics, thermoplastics and thermosets. Thermoplastics are materials that can be heated, softened, reformed, and cooled to a solid state many times. This is often compared to candle wax. Thermosets use heat (internal or external) to cure the material and once cured, they cannot be softened and re-formed. A common, non-polymer, example is concrete which goes through a chemical reaction generating internal heat and curing the material. Concrete cannot be ground up and re-cast. About 80% to 90% of the plastics used are thermoplastics and these are the materials used in the thermoforming process.

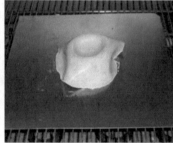

Figure 1.13: **Figure 1.14:**

Figure 1.15:

Figure 2.3: Paraffin - Thermoplastic

Figure 2.4: Concrete - Thermoset

2.2   MOLECULAR STRUCTURE

When polymers are formed, there are three common molecular structures, cross-linked, linear and branched. Cross linking occurs when a thermoset material polymerizes, or cures, and the molecules link together in a network or web often compared to a spider's web. (Fig. 2.5) Thermosets cannot practically be thermoformed since the sheet must first be created, thus the sheet would be a thermoset could not be heated and re-formed.

A linear molecule is something like a rope with very few side branches. They can be packed tightly together without the interference of side branches. Consider stacking bamboo sticks side by side. (Fig. 2.6) This stack would be very dense. On the other hand, branched molecules have many side branches creating a three-dimensional structure. Consider stacking many oak branches together. (Fig. 2.7) The three-dimensional structure of the branches holds them apart creating a much less dense structure. Using polyethylene as an example, chemist's can produce linear and branched material. The linear material tends to be higher density, less translucent, harder, and stiffer and have better chemical resistance than the lower density branched material.

Figure 2.5: Spider Web - Cross Linked

Figure 2.6: Bamboo - Linear

Figure 2.7: Oak - Branched

2.3   MORPHOLOGY

Thermoplastics are further characterized by their degree of crystallinity. When discussing plastics, the term crystalline refers to a very orderly grouping of molecules rather than a true crystal structure. Amorphous materials lack these orderly regions. In fact, the term amorphous means "without order." Consider spaghetti, a handful of the uncooked pasta (Fig. 2.8) is quite orderly while a bowl of cooked pasta (Fig. 2.9) has no order and the arrangement of the pasta is quite random.

Figure 2.8: Orderly Structure   **Figure 2.9:** Amorphus - Without Order

Crystalline and amorphous materials have quite different characteristics. In addition, the degree or amount of crystalline areas present in a polymer affects these characteristics. Crystalline materials tend to be translucent rather than transparent because the crystal regions are denser and do not allow light to penetrate easily. For example, low density polyethylene (LDPE) is about 65% crystalline while high density polyethylene (HDPE) is up to 95% crystalline. HDPE is harder, denser, has better chemical resistance, a higher melt temperature and is less translucent than LDPE. Other common crystalline materials include: polypropylene, nylon, and PET. Amorphous materials, those without crystalline regions, often have large side groups such as polystyrene which includes a benzene ring in the styrene monomer. These side groups prevent the formation of crystalline areas. As a result, many amorphous materials including polystyrene, polyvinyl chloride, and acrylic are transparent as there are no crystalline regions distorting the light.

The degree of crystallinity is quite important to the thermoforming process. Amorphous materials change gradually when heated. They slowly soften as more heat is applied until they have the characteristics of a very viscous liquid. The more heat that is applied, the more pliable the material and the easier it is to form or stretch. This "wide softening range," sometimes spanning over 100 degrees Fahrenheit, gives the thermoformer a lot of processing opportunities. For example, a shallow part may require much less heat to form, thus saving time, energy and therefore money. On the other hand, a crystalline material has a sharp melting point (melt temperature Tm) and until the material is within about 5 – 10 degrees of that temperature, no forming can occur. During the thermoforming process the melt temperature must not be reached or the sheet will rip or fail in the oven. Therefore, tight process control must be in place for the small forming window to be effectively controlled.

In fact, just before the melt temperature is reached, the crystalline areas break down and become amorphous. This is called the glass transition temperature or Tg.

2.4   VAN DER WAALS' FORCES

The thermoplastics molecule is made of a very strong carbon backbone held together most commonly with covalent bonds. These very long, macromolecules, determine the primary properties of the plastics material. These molecules are held in position to create the actual part by secondary bonds known as Van der Waals' forces. These are about 200 times weaker than the primary covalent bonds. The weaker attraction allows plastics molecules to slide by each other when the materials are heated without breaking the primary bond.

2.4.1   DISCOVERING VAN DER WAALS' FORCES

The following activity is an excellent method to understand these two different bonds:

1. Obtain a polyethylene "six pack" can holder.

2. Place two fingers from each hand in a single can ring and pull rapidly; it is not necessary to break the ring. (Figs. 2.10 and 2.11)

Figure 2.10:   Figure 2.11:

3. Note the area of elongation! It is deformed and is thinner than the un-stretched area. (Fig. 2.12)

4. Cut this elongated area across the narrow direction. (Fig. 2.13)

5. Using your fingernails, separate the molecules along the elongated area until the resistance increases. (Figs. 2.14–2.15)

6. Results: When the polyethylene was stretched (Fig. 2.17), the molecules overcame the Van der Waals' forces and aligned themselves in the order of the tensile pull. This demonstrated a phenomenon known as molecular orientation. When the part was cut and separated in the elongated direction, the weaker Van der Waals' forces were easy to separate in the oriented area. (Fig. 2.18) When the separation reached the non-oriented area, the materials seemed stronger as the covalent bonds must be broken to continue the separation.

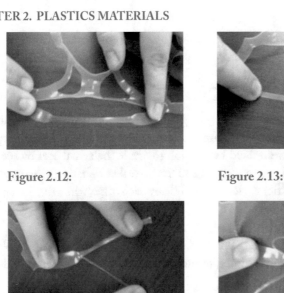

Figure 2.12:

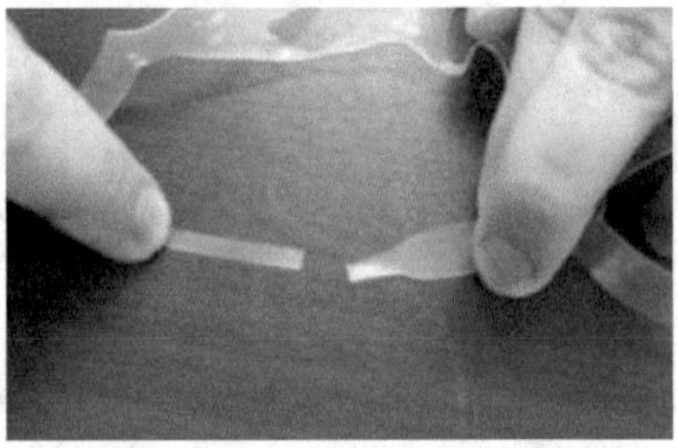

Figure 2.13:

Figure 2.14:

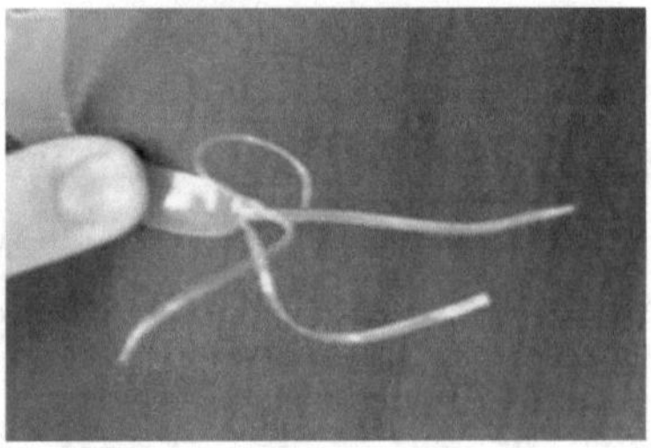

Figure 2.15:

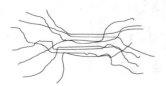

Figure 2.16: Polyethylene

Figure 2.17: Stretching Polyethylene

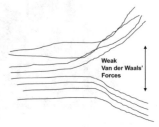

Figure 2.18: Separating the Molecules

2.5 MOST USED MATERIALS

Although almost any thermoplastic that can be created into a sheet form can be thermoformed, there are materials that dominate this industry. The following are the most used materials in the thermoforming industry, some key properties as well as common end products.

2.5.1 POLYSTYRENE (PS)

General purpose polystyrene, also known as GPPS is one of the easiest materials to thermoform with a very wide softening range. It is a transparent amorphous material and may well be the most used material in the thermoforming industry. Although optically clear, polystyrene scratches easily and has poor weatherabilty limiting its use to interior applications. It is also brittle and fails easily under impact. The low cost and ease of forming, however, make its use very high particularly in the disposables and packaging industry. In addition, thin expanded polystyrene (Styrofoam) can also be formed. Common thermoforming applications include disposable plates and cups. Packaging such as cottage cheese containers and blister packs are also common. There are industrial products made from polystyrene including refrigerator and freezer liners and door panels. To overcome the low impact strength of GPPS, butadiene may be added creating a material with higher impact while maintaining low cost and excellent forming properties. This material is known as high impact polystyrene or HIPS.

2.5.2   ACRYLONITRILE BUTADIENE STYRENE (ABS)
This "ter" polymer combines the properties of its three components: acrylic, butadiene (rubber), and polystyrene. The acrylic gives high gloss and good weatherabilty. The rubber increases impact strength and the polystyrene improves formability. This opaque amorphous material is typically used where high impact strength is needed yet clarity is not required. Common thermoforming applications include suitcase shells and housings.

2.5.3   POLYVINYL CHLORIDE (PVC)
PVC is hard and rigid. Impact modifiers must be added to increase toughness. PVC is self extinguishing making it ideal for transportation and building applications where flammability ratings may be an issue. PVC also has excellent weatherabilty properties which also makes it a good choice for exterior applications. In the packaging industry, PVC is being replaced with PET due primarily to the demand for recycled PET. Highly plasticized PVC is soft and flexible and is often used as a coating material to give automotive interiors a soft, textured feel.

2.5.4   POLYMETHYL METHACRYLATE (PMMA OR ACRYLIC)
This amorphous, transparent material is best known for its clarity (about 92%). It has good weatherability, UV resistance, high rigidity and good impact strength and has grades that are easy to thermoform. Common thermoforming applications include: signage (most backlit signs), skylights, formed aircraft windshields and hot tub shells which are typically coated with fiberglass reinforcement on the underside. Common trade names include Plexiglas and Lucite.

2.5.5   HIGH DENSITY POLYETHYLENE (HDPE)
This low cost crystalline material has very low water vapor permeability but relatively high gas permeability. These characteristics can be used to determine appropriateness in particular packaging applications for food products. HDPE also has excellent resistance to chemical degradation and is high impact and has good weatherabilty. Common applications include food packaging, waste containers, marine applications, pallets and tote boxes.

2.5.6   LOW DENSITY POLYETHYLENE (LDPE)

Having lower density than HDPE, this material is much less rigid and is used in flexible packaging.

2.5.7   POLYPROPYLENE (PP)

This crystalline material is known for its high strength and stiffness, chemical and moisture resistance and good impact strength. It is also resistant to tearing making it excellent for applications where continued flexing is required such as "living hinges." Additionally, PP can be sterilized. PP also has very low water vapor permeability but relatively high gas permeability Common applications include food and medical packaging, medical equipment housings and suitcase shells.

2.5.8   CELLULOSICS

This family of polymers is derived from plant fiber rather than petroleum. Two materials in this family are cellulose acetate (CA) and cellulose acetate butyrate (CAB). These materials do not attract dust because of their low electrostatic chargeability. CAB has higher impact strength than CA as well as excellent weatherabilty. These materials are most commonly used for blister packaging but may be used for signage and dust covers.

2.5.9   POLYETHYLENE TEREPHTHALATE (PET)

This material is also known as thermoplastic or saturated polyester. Although a crystalline material, PET can be processed to produce clarity. It has excellent barrier properties against gases, as well as moisture, making it a excellent packaging material and has been replacing other materials in the packaging and disposables sector due to the high demand for recycled PET.

2.5.10   GREEN PLASTICS

There is a movement away from petroleum based plastics as well as increases the use of degradable materials. This is particularly true in the disposables and packaging sectors of the thermoforming industry. More will be discussed in Chapter 9 called Environmental Issues.

2.6   IMPORTANT TERMINOLOGY

2.6.1   HEAT DEFLECTION TEMPERATURE (HDT)

This is actually the result of an ASTM test (D648) and an ISO test (75-1 & 75-2). Fig. 2.19 shows the three point deflection of the plastic test bar which is made to exacting specifications per the standard. After the bar is placed on the two lower supports and heated in a bath of oil. The dead weight is placed in the center of the bar and the oil is heated at a specific rate which in turn slowly heats the plastic bar. As the plastic increases in temperature, it begins to soften. The softening properties vary between plastic types and particularly between amorphous (wide softening range) and crystalline (sharp melting point) materials. This is one of the oldest tests used in the plastics industry, well before it was standardized by ASTM. For the thermoforming industry, this test can determine the lower forming temperature for materials to be thermoformed.

2.6.2   GLASS TRANSITION TEMPERATURE (TG)

This is the temperature above which thermoplastic materials (particularly relevant with amorphous materials) become rubbery rather than "glassy" or rigid. Recall that amorphous materials have a wide softening range so the Tg would relate to the lowest temperature at which a rubbery characteristic appears. Polystyrene and acrylic have a Tg of about 210 F and rigid PVC is about 185 F. Crystalline materials such as polyethylene, polypropylene and nylon have very low Tg's making this measurement of little value to the thermoformer. Tg for nylon 6, for example, is about 120 F, polypropylene is about 15 F, and polyethylene is about -125 F.

2.6.3   MELT TEMPERATURE (TM)

It represents the melting temperature for crystalline materials. Recall that crystalline materials have a sharp melting point which is critical for the thermoformer to know as this process must not reach the materials melt temperature, however the forming temperature may be only a few degrees below Tm.

2.6.4   POLYMER SET TEMPERATURE

This is the temperature below which the part can be removed from the mold without distortion.

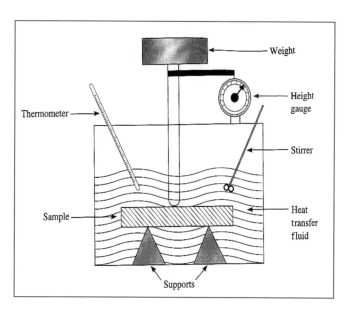

Figure 2.19: Heat Deflection Test

2.6.5 HOMOPOLYMER

is a plastic material made entirely from one type of material or monomer. When two plastics are combined, it is called a copolymer. Plastics are typically combined to modify properties or characteristics. For example, if polystyrene and butadiene rubber are combined the impact strength is increased and the new copolymer is called High Impact Polystyrene or HIPS. Polyvinyl Chloride may be combined with Polymethyl Methacrylate to form the copolymer PVC-PMMA. Three materials may also be combined to create a ter-polymer such as ABS which combines acrylonitrile, butadiene and styrene.

2.6.6 THERMOFORMING WINDOW

This is the temperature range in which a particular material may be formed. For an amorphous material, this window may be over 100 F and may be specified as the lower forming temperature, normal forming temperature and upper forming temperature. Crystalline materials have a window of only a few degrees as the crystalline regions break down and the material may be formed. Polypropylene, for example, has a forming window of only two or three degrees just below its Tm.

CHAPTER 3

Thermoforming Process Overview

3.1 THE THERMOFORMING PROCESS

The thermoforming process is considered a secondary process in the plastics industry as the input materials, plastics sheet and film, must first be created. This sheet and film are typically produced using extrusion, casting and calendaring processes. The fundamental thermoforming process requires the following steps: sheet *preparation*, *loading* the sheet into the process, *heating* the sheet to its forming temperature, *stretched* the sheet into the desired shape using some type of force, *cooling* the sheet to a temperature where the new shape will be sustained, *unloading* the part from the process, and *trimming* the part to its desired final form. The following chapter will discuss each of these process steps.

There are three basic types of thermoforming machines: cut sheet shuttle machines, cut sheet rotary machines, and roll feed continuous machines. The process steps vary within these machine types, although all steps are includes within each process. The shuttle machine, illustrated in Figs. 3.1–3.3, handles a single cut sheet at a time, thus this is a slow process commonly used for low volume or very large products.

The rotary machine has multiple clamping frames for cut sheet, typically 3 or 4. One station is for loading and unloading. Another station contains the heating system and the other station contains the forming and cooling processes. Fig. 3.4 shows a typical three station rotary machine.

The continuous thermoforming machine incorporates many, and possibly all, of the process steps in a single machine. Within this machine, roll feed material is heated, formed, cooled, trimmed and often stacked for packaging. This process is used most often for very high volume products including many disposables as well as packaging. Fig. 3.5 shows a continuous machine in action.

3.2 SHEET PREPARATION

The amount and type of sheet preparation depends on the sheet thickness as well as the process being used to form the sheet. Sheet thickness is classified by gauge. Thin gauge is typically less than .060 in. (1.50 mm) in thickness, film or foil is less than .010 in. (0.25 mm). Thick gauge is larger than .120 in. (3.0 mm) and material thicker than .5 in. (13 mm) is typically called plate. Thin gauge and film are handled in rolls as this material is flexible. In addition other flexible materials, such as polystyrene foam, of greater thickness may also be rolled. Mid-range materials, .060 in. (1.50 mm) to .120 in. (3.0 mm) and thicker are typically handled as cut sheets and stacked flat to eliminate problems with curling or other undesirable deformation prior to processing.

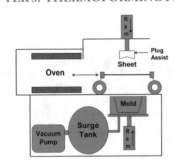

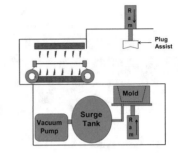

Figure 3.1: Shuttle Machine   **Figure 3.2:** Heating Process

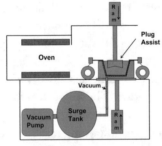

Figure 3.3: Forming Process

Figure 3.4: Typical 3 Station Machine

There is little preparation for rolled materials as the sheet or film is fed directly into the machine from the rolls. This process is known as roll fed thermoforming and is typically highly automated and continuous. This in-line process includes all steps listed above in a single process and is used for very high volume products such as disposable plates, drinking cups and some packaging.

Figure 3.5: Continuous Former & Trim Press *Courtesy Brown Machine LLC.*

Sheet preparation and loading of cut sheet materials includes cutting or sheering the sheets to the proper size needed for the clamping unit to hold the sheet during processing. The type of equipment used with cut sheet thermoforming includes single station or shuttle machines and rotary machines. The single station machine is typically used for lower volume products and prototypes when cycle times are not as critical. In this type of machine only one sheet is used at a time and the cycle time includes loading, heating, forming, cooling, and unloading in a sequence. Trimming is normally performed as a secondary operation. Rotary machines are used when higher volumes are required and cycle time is more critical.

3.2.1 DRYING

Some materials are hygroscopic which means they absorb moisture at the molecular level. If moisture is present during the heating process, it turns to vapor and creates quality flaws in the product. Materials such as polycarbonate, acrylic, and nylon may require extensive drying if the sheet has been exposed to humid environments during storage. This adds to the cost of the products. If the sheet material is not stored properly or is stored for extended periods of time, it will also absorb moisture. The best practice is to tightly seal these polymer sheets in a moisture barrier such as polyethylene film at the extrusion process. If drying is required, it is performed in recirculation ovens at about 150-300 degrees F and may take between 2 and 4 hours depending on the material and the sheet thickness. Specific information for drying can be provided by the material manufacturer.

3.3   LOADING

The loading process also varies by material thickness and the forming equipment to be used. Thin gage materials are typically "roll fed" using a continuous process. Pre-cut sheets of thicker gage materials are loaded manually or automatically into "sheet fed" machines. The cut-sheets are held in a clamping frame on all edges so when heated they will not twist or warp. (Fig. 3.6) This frame is also a key part of the forming process to be discussed later.

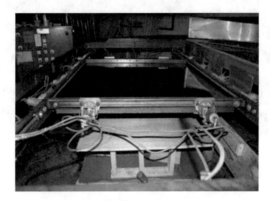

Figure 3.6: Sheet Held in Clamping Frame
Courtesy of Helton Inc.

3.4   HEATING

The purpose of the heating process is to heat the sheet evenly throughout. Before discussing how this is accomplished in the thermoforming process, this section will review the three methods of heat transfer between two surfaces. These three methods are known as radiation, conduction and convection and all three are used in the thermoforming process. (Fig. 3.7) Radiation is primarily used to heat the sheet to the desired temperature. Conduction is used to heat the core of the sheet as well as in the cooling process. Convection is used primarily to cool the part so it can be removed from the mold quickly; however, forced air convection ovens may be used when heating very thick material to reduce surface scorching.

The thermoforming process heats the surface of the sheet material using radiation which is the interchange of electromagnetic energy between solid surfaces of different temperatures. Common heat sources used in this process include combustion heaters (natural gas), calrod heaters, ceramic heaters (Fig. 3.8), quartz heaters (Fig. 3.9), etc.

Each of these has particular advantages and disadvantages including original cost, operating cost, robustness, heat control accuracy, etc. Individual equipment manufacturers can provided specifics and help determine the best option for a particular application.

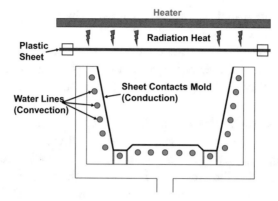

Figure 3.7: Heat Transfer in Thermoforming

Figure 3.8: Multi-Zone Ceramic Heaters
Courtesy of Brown Machine LLC.

Figure 3.9: Multi-Zone Quartz Heaters
Courtesy of Brown Machine LLC.

Plastics are natural thermal insulators. It is very difficult to transfer heat into the center of a thick sheet of plastic. For example, the thermal conductivity of aluminum is almost 700 times that of polystyrene. Radiant energy is absorbed into the surface of the sheet to a depth of approximately 0.010" – 0.030." Conduction must then transfer the heat into the core of the sheet. Thermal conductivity is much less important when forming thin gage versus heavy gage materials which may scorch on the surface before the core reaches forming temperature. As a result, thermoforming has a natural limit on sheet thickness. Forming sheets above 0.500" in thickness require special expertise and equipment.

There are several variables which must be considered regardless of the heat source to be used. One such variable is the type of material to be formed, as different plastics require different amounts of radiant energy to reach forming temperature. Crystalline plastics require more energy input to increase in temperature than do amorphous materials. This varies further with the type and amount of fillers, reinforcements and even the color of the sheet.

The space between the sheet and the heat source is another important variable as heat is lost as the space increases. Heater element spacing creates yet another variable. For example, if calrod heaters are used and they are 4" apart, there will be natural variation in sheet temperature based on this spacing.

As plastics are natural thermal insulators, often the heating step is the longest step in the process. This can be reduced by modifying the process such as creating two heating steps. A typical sheet fed rotary machine has three stations including 1) Load/Unload, 2) Heat, 3) Form/Cool. (Fig. 3.10) In this process the cycle time is driven by the time required to heat the sheet to the correct forming temperature. One method to significantly reduce the cycle time is to add a fourth station making this a four station machine with two heat stations. (Fig. 3.11) This has the potential to cut the cycle time in half.

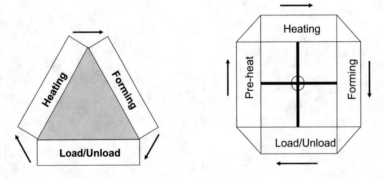

Figure 3.10: Three Station Machine   Figure 3.11: Four Station Machine

3.5   FORMING

Recall that thermoforming is a stretching process where properly heated thermoplastic sheet is stretched into a new shape using some type of force. There are typically four options regarding the force available to form the sheet. 1) *Mechanical force* can be applied to stretch the sheet. This may be as simple as a strip heater softening a limited area of the sheet and manually bending the sheet to form a new geometry. Mechanical forces can be quite high. 2) The most common force used in thermoforming is *atmospheric pressure* which pushed the sheet into a new shape after a vacuum is drawn between the sheet and the mold. This force is quite limited with a maximum force of approximately 15 PSI. This vacuum forming process (Fig. 3.12) is widely used for high volume thin gage forming where only minimal pressure is required to stretch the sheet to the desired shape. Fig. 3.13 shows a large vacuum forming mold for a truck bed liner. 3) When atmospheric pressure is inadequate to form the sheet to the desired shape or the desired detail, *pressure forming* may be used.

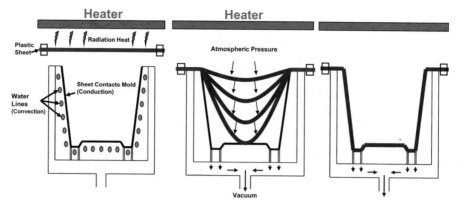

Figure 3.12: Basic Vacuum Forming Process

Figure 3.13: Vacuum Forming Mold for Truck Bed Liner *Courtesy Tooling Tech. Group*

In this case, compressed air rather than atmospheric pressure is used which has a practical limit of about 150 PSI or 10 times the force of vacuum forming. 4) Large and detailed parts often incorporate a *combination of forces* to form the sheet to the desired geometry. For example, a mechanical force may stretch the sheet followed by a vacuum to draw the sheet into the mold. This is often used to reduce the wall thickness variation. The forming step will be discussed in more detail in the next chapter.

3.6   COOLING

Cooling begins immediately when the sheet contacts the mold surface. Using the concept of conduction heating, the hot plastic material is heating the cold mold. The quicker the part can be cooled, the faster it can be removed from the mold. Aluminum is the most common mold material for high volume parts and for virtually all roll-fed processes. The aluminum mold is cooled by passing temperature controlled water through drilled passages. Using the convection heating concept, the hot mold is heating the cold water. Addition convection cooling is often incorporated by blowing air or a fine water mist across the formed part. The cooling process can be very quick or painstakingly long depending on the mold material and part thickness. A roll-fed, thin gage, part cooled on an aluminum mold that is just above the temperature where condensation occurs may be as little as 1 second. Cooling a thick gage sheet of acrylic on an epoxy tool may take several minutes. Epoxy and other mold materials will be discussed in Chapter 6, Tool Design.

As with the heating process, there are many variables in the cooling process which affect the total cooling time. These include the type and grade of the plastic being cooled, the material thickness after stretching, the material temperature, the polymer set temperature, the mold temperature, the mold material, etc.

3.7   UNLOADING

The key issue with unloading a part from the machine is timing. If the part is not sufficiently cooled, it may deform after removing. If it is too cold, the cycle is too long and profits suffer. Cavity (female or negative) and plug (male or positive) molds have different issues when unloading. As plastic material cools, it shrinks. Therefore, the part shrinks away from the cavity mold making unloading easy. (Fig. 3.14) However, the part shrinks onto the plug mold creating a tight fit which can be difficult to release. (Fig. 3.15) Mechanical or air ejection can help as well as providing the maximum draft angle allowable for the product. Plug and cavity design issues will be discussed further in Chapter 5, Part Design.

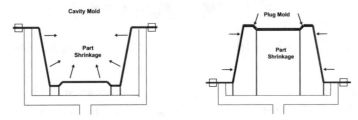

Figure 3.14: Cavity Mold Design   **Figure 3.15:** Plug Mold Design

Roll fed machines automatically unload the part as the material moves through the process. Sheet fed machines require manual or automated part removal. This may require several operators for large products such as bath tubs, burial vault liners and large skylights. Robotics can be effectively used when the volume can justify the costs. Another key issue with unloading is part quality. Packaging for cosmetics may require much more care than packaging for a disposable razor. Unloading an acrylic skylight can be quite touchy as the material scratches easily and the customer expects a defect free product.

3.8   TRIMMING

The plastic part must be held in a frame during the thermoforming process. Therefore, all thermo-formed parts must be cut out of the formed sheet. This can be quite simple, such as a straight sheer, or very complex requiring a computer guided laser. The trimming method to be used is dependent on many factors including:

Material Type – some material is hard and brittle while others are soft and tough.
Material Thickness and Variation
Complexity of the Part – Planer versus non-planer trimming
Dimensional Tolerances
Cosmetic Requirements
Other Operations – drilling, punching, etc.

Trimming can be broken into several categories. Two such categories are sheering and chip removal processes. When material is sheared, no material is lost and the sum of the parts equals the whole. For example, cutting paper with a scissors is a shearing process as there was no paper lost in the process. Shearing processes are very clean and create no dust or chips. During the chip removal processes, there is a loss of material in the form of dust particles, strands, chips or vapors in the case of laser trimming.

Sheering processes include punch & die sets, steel rule dies and in mold trimming. These may be in the mold itself, (Figs. 3.16, 3.17, & 3.18) in line with the machine, or separated as a secondary operation. The blades may be heated to aid in cutting and improve the quality of the trimmed edge.

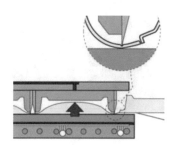

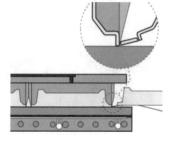

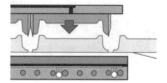

Figure 3.16: Form Part
Courtesy GN Thermoforming

Figure 3.17: Trim Part
Courtesy GN Thermoforming

Figure 3.18: Remove Part
Courtesy GN Thermoforming

In mold and in line sheering processes are typically used for thin gage high volume products such as packaging and disposables. The process may also include counting, stacking and packing of the products. Sheering processes do not work well on hard or brittle materials such as acrylic or thick general purpose polystyrene. These materials will fracture or crack during this process.

Chip removal processes are used for thick gage and lower volume products. These processes include sawing, routing, drilling, water jet and laser cutting. The parts are normally located on a holding fixture. To improve accuracy and repeatability, these tools may be controlled by a CNC multi-axis system or attached to a robotic arm. Special blades, drills and routers have been designed for different plastics to optimize the speed and quality of these processes. Chip removal processes work well on most materials and are very effective on hard, brittle materials. However, these processes are quite slow compared to the rapid rate of shearing processes. Figs. 3.19–3.24 show a variety of chip removal examples.

Figure 3.19: Band Saw

Figure 3.20: Simple Table Saw Fixture

Figure 3.21: Tracer Router

Figure   3.22: CNC   Routing/Trimming
Courtesy of Helton Inc.

Figure 3.23: Simple Drill Fixture
Courtesy of Helton Inc.

Figure 3.24: Routing Fixture
Courtesy of Helton Inc.

CHAPTER 4

The Forming Process

4.1 FORMING INTRODUCTION

Forming plastics during the thermoforming process is basically a stress/strain issue. As the material is heated, the amount of force, stress, needed to deform the part, strain, is reduced. Therefore, the stress/strain curve varies greatly with temperature. When thermoplastic materials are within their thermoforming temperature window, the amount of stress required for deformation can be quite low, as little as 2 PSI for polystyrene. As the length of deformation required increases or the level of detail in the part increases, the amount of force must increase. However, this force or stress rarely reaches 200 PSI. For these reasons, thermoforming is considered a low pressure process.

As discussed in Chapter 3, there are four types of forces used to form the sheet material during this process. These are: mechanical, atmospheric pressure, compressed air, and a combination of these forces. Each of these will be discussed further in this chapter.

4.2 MECHANICAL FORMING

Mechanical forming of plastics spans the continuum from very simple to very complex and expensive. Some of the simplest forming processes are mechanical. Using a strip heater to heat a select area of a sheet and simply bending the sheet to a new shape is one example of mechanical forming. (Fig. 4.1) This process is used for low volume items such a drink menu holders in restaurants, picture holders, and various retail displays. (Figs. 4.2–4.4)

Figure 4.1: Simple Strip Heating & Mechanical Forming Example: Heat - Form Hold/Cool

On the other end of the complexity, and cost, continuum is matched mold forming. In this process two molds are created with space between them for the heated sheet. The mold closes onto the sheet, forming it to the mold geometry where it is cooled. This process can produce the greatest forming force and can add details to the part surface such as lettering, embossing, textures, etc. This process is very similar to the metal stamping process used to produce body parts for automobiles. In

Figure 4.2: Formed & Fabricated Display Case

Figure 4.3: Formed Menu Holder

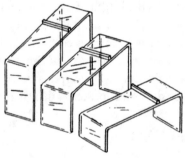

Figure 4.4: Formed Shoe Display

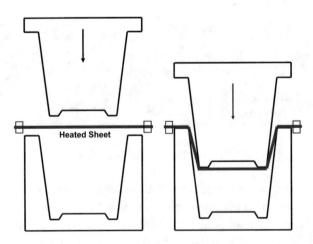

Figure 4.5: Matched Mold Forming

mold trimming is also frequently done in this process as the forces required for trimming even heavy gage materials are used to form the sheet. This process is the most costly of the thermoforming processes and therefore has limited application.

4.3   VACUUM FORMING

Straight vacuum forming utilizes atmospheric pressure to force the heated sheet against the mold surface where it cools. Although this force is quite limited, about 15 PSI maximum, this is the most common process used for high volume thin gage products. In this process the heated sheet is placed over a cavity mold. Contact is made between the sheet and the mold creating a seal. The air in the cavity is evacuated and atmospheric pressure forces the sheet against the contours of the cavity. Most vacuum forming machines include a surge tank which is first evacuated so the forming can occur very quickly in the process.

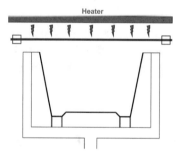 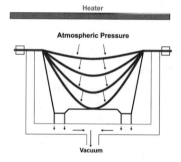

Figure 4.6: Heating Sheet   Figure 4.7: Forming Sheet

4.4   PRESSURE FORMING

Pressure forming utilizes compresses air to force the heated sheet into the contours of the mold cavity. This force can easily exceed ten times that of straight vacuum forming thus allowing more detail and further stretching. This process is similar to straight vacuum forming as the heated sheet (Fig. 4.8) is sealed against the mold; however a pressure box is then sealed over the sheet. Compressed air is injected into the pressure box pushing the sheet into the mold cavity and against its contours. The air between the sheet and he cavity must be vented out of the cavity. (Fig. 4.9) Often vacuum is applied to expedite this process. Since the force is much greater than vacuum forming, letters, embossing, sharper corners, etc. can be formed. Of course these details are only on one side of the part unlike matched mold forming which can apply details to both sides. The molds must be more robust for the pressure forming process than the straight vacuum forming process as the forces against the mold are greater. This process is typically applied to form thick gage materials, products requiring extreme detail, and hard to form materials such as polycarbonate.

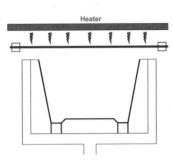

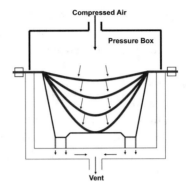

Figure 4.8: Heating Sheet   Figure 4.9: Forming Sheet

4.5   COMBINATION FORMING PROCESSES

Often none of the three processes discussed previously may be ideal for all applications. Therefore, combination processes have been developed to achieve the benefits of two or more of the above to best meet a particular need. This may be done to reduce wall thickness variation, use thinner material thus reducing costs, reduce cycle time, etc.

4.5.1   DRAPE FORMING

Drape forming combines the mechanical forming process with the forces of vacuum forming. A plug or positive mold is used rather than a cavity mold. The mold is mechanically pushed into the heated sheet thus stretching it until the base of the plug creates a seal. (Fig. 4.12) The space between

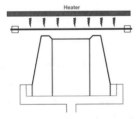

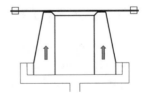

Figure 4.10: Heating Sheet   Figure 4.11: Mold Rises

the plug and the sheet is then evacuated and atmospheric pressure forces the sheet against the plug. (Fig. 4.13) As noted in Chapter 3, the part will shrink on to the plug making the part difficult to remove. Often compressed air is blown through the vacuum holes to help release the part. If more detail is needed than can achieved with atmospheric pressure, a pressure box can be placed over the plug and compressed air can be used to form the sheet. Draft angles are critical on plug molds and this will be discussed in Chapter 5, Part Design.

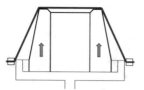

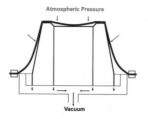

Figure 4.12: Mechanical Stretch

Figure 4.13: Atmospheric Pressure Forms Details

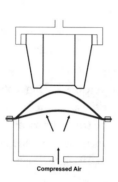

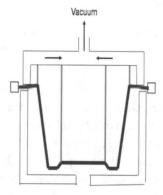

Figure 4.14: Billow Pre-Stretches Sheet

Figure 4.15: Plug Inverts Billow, Vacuum Applied

4.5.2   PNEUMATIC PREFORMING

Pneumatic preforming includes a variety of processes in which air is used to pre-stretch the sheet into a bubble known as a billow. As the billow is blown, as seen in Fig. 4.14, a plug mold is pushed into the billow and the sheet forms around the plug. Vacuum is then applied between the sheet and the plug to form the details. If more force is needed to form the sheet, compressed air may be applied to the sheet. An alternate version of this process involves pre-stretching the sheet into a vacuum box (Fig. 4.16), then pushing the plug into the reverse billow. Vacuum is then applied forming the sheet on the plug. (Fig. 4.18) This process is sometimes referred to as vacuum snapback. There are many other processes that can be created using pneumatics to pre-stretch the sheet. These are only limited by the process designer's imagination.

4.5.3   MECHANICAL PREFORMING

Mechanical preforming processes typically us a plug or positive device to pre-stretch the heated sheet before forming. These processes are known as plug assist and give the process designer the ability to pre-stretch some areas of the sheet more than other areas thus better controlling the wall thickness

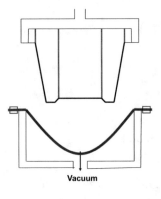

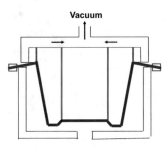

Figure 4.16: Vacuum Pre-Stretches Sheet

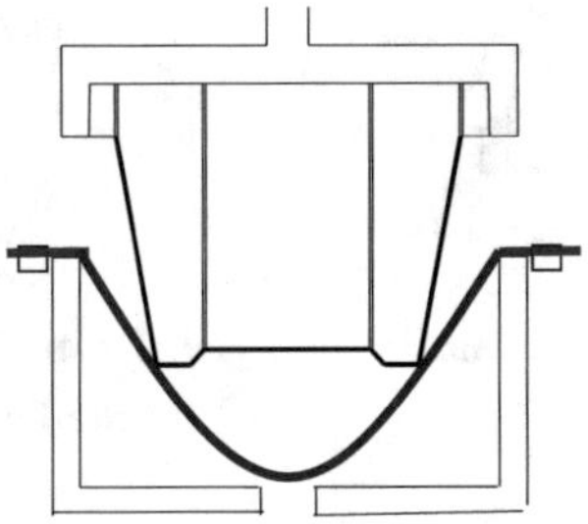

Figure 4.17: Plug Mold Inserted Reverse Billow

Figure 4.18: Vacuum Applied Plug-Sheet "Snaps-Back" Onto Mold

on the finished part. Disposable drinking cups are typically produced using a plug assist process. This produces a product with a thick bottom and thin walls. Plugs must be carefully designed and are often temperature controlled. The plug may be chilled to freeze the contact area or heated to allow further stretching during the forming step. As with pneumatic preforming, only the designers' imagination limits the use of mechanical preforming possibilities. Figs. 4.19–4.21 illustrate this process.

4.6   TWIN SHEET THERMOFORMING

Twin sheet thermoforming is a special process in which two thermoformed parts are bonded together during the process to form a hollow part. This process requires the use of two cavity molds and two heating stations. Both sheets are heated simultaneously and transferred to the forming station. Typical straight vacuum forming procedures are used to form both sheets into their respective cavities. (Fig 4.22) After forming, but while the materials are still hot, the two cavity molds are forced together thus welding the two parts into one. (Fig. 4.23) A needle may be inserted to blow air into the hollow part to further form the parts and provide more forming pressure. Foam may also be injected into the hollow part to make a foam filled structural part such as a very tough plastic pallet. The "pinch off area" design is very important to obtain a solid weld between the two halves.

4.7   LAMINATING THERMOFORMING

The thermoforming process may be used to shape and laminate two materials together. This may also be a covering process. In this process a part to be covered is used rather than a mold. The sheet material, often vinyl foam or decorative thermoplastic fabric, is heated and formed directly onto a part to be covered or laminated. The part may have been created by thermoforming, injection molding, and composite processing or may even be a metal part. This process has been used in the

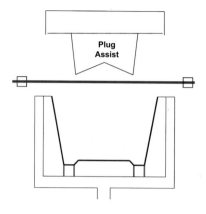

Figure 4.19: Heated Sheet

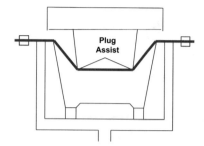

Figure 4.20: Plug Pre-Stretches Sheet

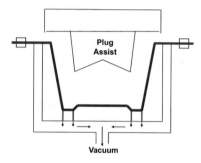

Figure 4.21: Vacuum Applied

transportation industry to create a decorative and/or protective layer on a part for automotive, truck, aircraft, and public transportation vehicle interiors. The original part typically has a heat sensitive adhesive applied before the thermoforming process which is activated when the heated sheet makes contact during forming.

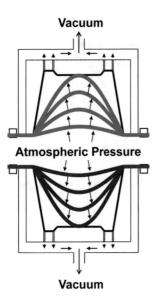

Figure 4.22: Forming Two Parts

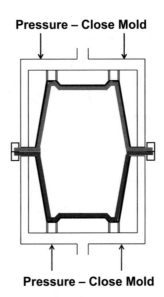

Figure 4.23: Molds Forced Together, Creates Weld

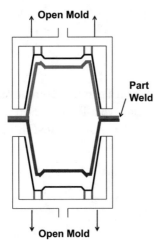

Figure 4.24: Welded Part Removed

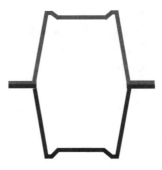

Figure 4.25: Completed Part

CHAPTER 5

Part Design

5.1 DESIGN QUESTIONS

Prior to discussing part design specifics, the first question to resolve is whether a part should be thermoformed. There are many sources for these questions as well as differing opinions as to their answers. The following represent a select set of design issues that need to be considered. Many of these will be further discussed in this chapter as well as the following chapter on tool design.

Will the part design allow for the #1 issue in thermoforming – wall thickness variation?

What material is to be used?

> Is the material available in sheet form?
>
> Is the draw ratio of this material adequate?
>
> Are there any types of fillers, additives or reinforcement required?

What is the actual part geometry?

> Corner radii?
>
> Draft angles?
>
> Depth of draw?
>
> Is webbing a concern?
>
> Are there any undercuts or negative draft angles?

What are the application requirements for this part?

> Useful temperature range?
>
> Strength requirements such as pressure, impact, stiffness, etc.?

What are the quality expectations?

> Cosmetic?
>
> Optical?
>
> Dimensional tolerances?

5.2   WALL THICKNESS VARIATION

The number one issue with thermoforming as a process is the fact that it is a stretching process therefore the wall thickness varies with the amount of stretching that occurs. In the previous chapter, several processing options were discussed to reduce the wall thickness variation in the finished part including pre-stretching the sheet with pneumatics or a mechanical plug. The part geometry has a huge impact on the amount of variation as does the decision whether to us a plug (male or positive) or cavity (female or negative) mold.

5.2.1   PLUG VERSUS CAVITY

A basic concept to keep in mind is that the material quickly cools as it touches the cold mold surface and stretching in that area ceases. The first area of the sheet to make contact with the mold will be the thickest and the last area of the sheet to touch the mold will stretch the furthest and therefore be the thinnest. Figure 5.1 shows an almost inverse relationship in wall thickness between parts of the same geometry when formed in a cavity type mold compared to those formed over a plug type mold. Both tool types are used in the thermoforming industry but each has an impact on the thinning location of material due to stretching.

Figure 5.1: Cavity Produced Part Compared to Plug Produced Part

A pair of molds was created to demonstrate the wall thickness differences between a part produced on a cavity mold and a plug mold. Figure 5.2 shows two molds to produce a small bowl. These molds are made of composites (fiberglass & polyester) and are mounted on a vacuum base which is compatible with the thermoforming equipment in use. The part geometry is nearly identical

Figure 5.2: Composite Cavity & Plug Mold

between mold types. The finished product is approximately 2.25" high and 6" in diameter. The thermoforming process in use requires a 12" X 12" sheet to be held in a clamping frame with a 10" X 10" area available for forming.

The two sheets to be formed in this test are marked with a $\frac{1}{2}$" X $\frac{1}{2}$" grid on each side prior to forming as shown in Fig. 5.3.

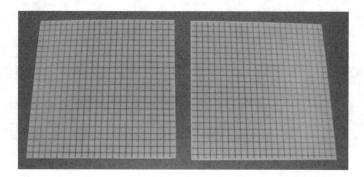

Figure 5.3: Test Sheets with 0.5" X 0.5" Grid

Cavity Mold Example

Figure 5.4 shows the cavity mold in position on the thermoformer. The wooded box beneath the mold is a vacuum box which is connected to a vacuum surge tank.

Figure 5.4: Cavity Mold on Machine

Figure 5.5 shows the heated sheet in place over the mold ready to be formed. Figure 5.6 shows the sheet being forced into the mold with atmospheric pressure. Note the grid pattern! The area of the sheet that first touches the mold has not deformed. The entire part was produced by the material within the 6" diameter cavity area. Figure 5.7 shows the part and the "trim;" again note the grid pattern and the degree of deformation.

Figure 5.5: Sheet Located in Clamping Frame

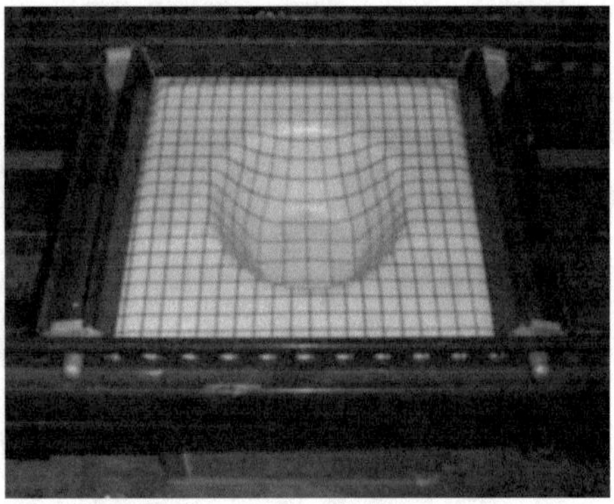

Figure 5.6: Heated Sheet Being Formed in Cavity Mold

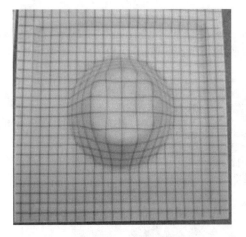

Figure 5.7: Formed Part with Trim

Figure 5.8: Plug Mold on Machine

Plug Mold Example

Figure 5.8 shows the plug configured mold in place on the thermoformer.

Figure 5.9 shows the plug mold being raised into the heated sheet mechanically stretching the material. This process is known as drape forming. Note the deformation on the grid! Figures 5.10 & 5.11 show the sheet being drawn over the plug mold where it will cool. Note the grid pattern and the area of deformation which extends to the clamping frame. Also note that the bottom of the

Figure 5.9: Sheet Located in Clamping Form

Figure 5.10: Sheet Being Formed on Plug Mold

Figure 5.11: Formed Part on Plug Mold

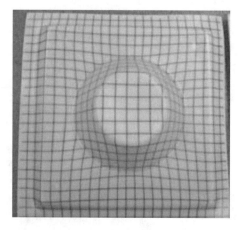

Figure 5.12: Formed Part with Trim

bowl (the area which first touched the mold) has almost no deformation, thus retaining its original sheet thickness. Figure 5.12 shows the part and "trim" after removal from the thermoformer.

Mold Design Comparison

Figure 5.13 & 5.14 show the plug and cavity formed units side by side for comparison. The deformation location and degree are quite obvious. Note the grid lines and the specific areas of greatest deformation between the plug molded part on the left and the cavity molded part on the right. The bottom of the cavity molded bowl is very thin due to extreme stretching while the bottom of the plug molded bowl is nearly unchanged from the original sheet thickness.

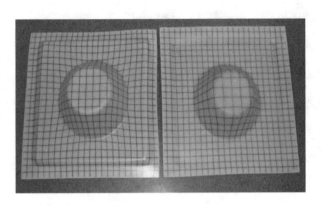
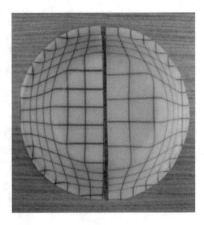

Figure 5.13: Plug Formed Part (left) and Cavity Part (right)   **Figure 5.14:** Plug Formed Part (left) and Cavity Formed Part (right)

Figure 5.15 shows wall thickness dimensions in .000" for the cavity mold. The starting sheet thickness was .150". The thinnest location of .060" represents 40% of the original thickness.

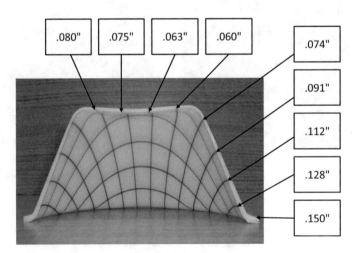

Figure 5.15: Cavity Formed Part Cross Section

Figure 5.16 shows wall thickness dimensions in .000" for the plug mold. The starting sheet thickness was .150". The thinnest location of .106" represents 70.6% of the original thickness.

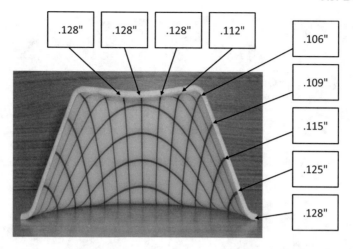

Figure 5.16: Plug Formed Part Cross Section

5.3   DRAW RATIOS

There are several Draw Ratios that can be used to analyze and compare parts. These include Aerial Draw Ratios, Linear Draw Ratios and Height-to-Dimension Ratios. Each has advantages but are only grossly representative of sheet thinning, however they can be excellent instructional tools for comparing part designs and processes. For illustration purposes, a pair of wooden test molds was used. These produce a simple rectangular box measuring 6.5" long X 2.75" wide X 2.75" deep. One mold produced a part using drape forming with a plug mold and the other produced a part using vacuum forming into a cavity. The clamping frame opening was 9.5" X 9.5". Both molds and parts are illustrated below in Figs. 5.17–5.20.

5.3.1   AERIAL DRAW RATIO (ADR)

ADR is the overall measurement of stretch of the sheet. This is determined by calculating the surface area of the formed part and dividing it by the surface area of the sheet used to form the part. This will always be a valued greater than 1.

ADR = Surface area of the formed section / Surface area of the sheet used to form the part

The aerial draw ratio can be used to determine the average reduction in sheet thickness by dividing it into 1.

Average Thickness Reduction = 1/ADR

Maximum ADR's are shown in Table 5.1. This information is helpful to compare the stretching properties of various materials.

Figure 5.17: Plug Mold (left) & Cavity Mold (right)

Figure 5.18: Plug Mold (left) & Cavity Mold (right)

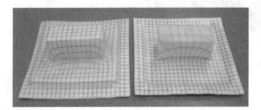

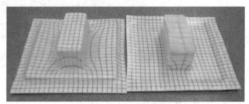

Figure 5.19: Plug Mold & Cavity Produced Parts

Figure 5.20: Plug and Cavity Produced Parts

Table 5.1: Aerial Draw Ratio (ADR)	
Plastic	Maximum ADR
ABS	5.5
Acrylic	3.4
HDPE	6.5
LDPE	6.0
PP	7.5
PS	8.0
PVC	4.3

Using the parts shown in Fig. 5.17, the part produced using a plug mold and the drape forming process would have an aerial draw ratio as follow:

The surface area of the part is 68.875 square inches. Since this is a drape formed part, the area of the sheet within the clamping frame must also be included. This would be 9.5 X 9.5 – (the area of the sheet which made the part 2.75 X 6.5) = 72.375. Therefore, the total area of the formed sheet is 68.875 + 72.375 = 141.25 square inches. The areal draw ratio for this part = 141.25/ (9.5 X 9.5 starting sheet size) = 1.565. In other words, the sheet had to stretch 1.565 times its original size to create this part using the drape forming process. The average thickness reduction is 1/1.565 or 63.8%.

Producing the same part using the vacuum forming process and a cavity mold has dramatically different results than the drape formed part. Straight vacuum forming produces the part from only the sheet is within the cavity opening. Therefore, the sheet must stretch further when using this process. The following illustrates the areal draw ratio for the rectangular part produced using a cavity mold:

The area of the part is 68.875 square inches. The area of the sheet used to produce the part is 2.75" X 6.5" = 17.875 square inches. Therefore, the areal draw ratio is 68.875/17.875 = 3.853. In other words, the sheet must stretch 3.853 times it original length to create this part using straight vacuum forming and a cavity mold. The average thickness reduction is 1/3.853 = 25.9%.

5.3.2   LINEAR DRAW RATIO (LDR)

This the comparison of the length of a straight line drawn on the sheet before forming as compared to the length of the same line after forming. Only the forming area is included in this calculation.

LDR = Line length on formed part / Line length before forming

If LDR is used as an evaluation tool, the part is typically assessed in the direction requiring maximum draw. Using the rectangular box described above, the part would be evaluated across the narrow dimension. As with areal draw ratios, there is a difference in calculation between parts made with cavity and plug molds.

Figure 5.21 illustrates the LDR parameters for a part produced using a plug mold and the drape forming process. The line length before forming is 9.5" and after forming is 15". Therefore, the LDR = 15/9 = 1.67 or the line is now 167% of its original length.

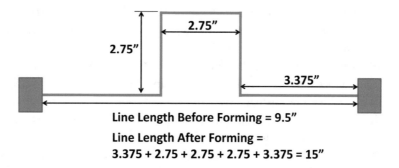

Figure 5.21: LDR for Plug Produced Part

Figure 5.22 illustrates the LDR parameters for a part produced using a cavity mold and the vacuum forming process. The line length before forming is 2.75" and after forming is 8.28". Therefore, the LDR = 8.25/2.75 = 3 or the line is now 300% of its original length.

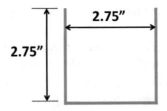

Line Length Before Forming = 2.75"

Line Length After Forming =
2.75 + 2.75 + 2.75 = 8.25"

Figure 5.22: LDR for Cavity Produced Part

5.3.3   HEIGHT – TO - DIMENSION RATIO

This ratio is simply the height of the formed part divided by the length of the greatest opening of the part. The usefulness of this ratio is limited to simple symmetric parts such as a drinking cup using straight vacuum forming process with a cavity mold.

H:D = Height of formed part / Greatest length of opening

H:D Example: The deep draw cup in Fig. 5.23 is 7" tall and the opening is 3.875. Therefore, the H:D is 7/3.875 or 1.806

5.4   MATERIAL SELECTION

The first question to ask is whether the material to be used is available in sheet form at the desired thickness. If it is not, are the volumes such that it could be made for this application economically? If this is not the case, are there other materials that are available that may meet the requirements?

Another question is whether the draw ratios of the selected material are adequate for this application? Thermoplastics vary greatly in their ability to be stretched without failing. Polystyrene for example has an aerial draw ratio of 8.0 meaning it can stretch 8 times its original volume without failing. Acrylic on the other hand has an aerial draw ratio of just 3.4. Although both of these materials are commonly thermoformed, polystyrene lends itself much better to deep draw parts.

If any types of fillers, additives or reinforcements are required in the material selected, the stretch characteristics will be affected. Additives are chemicals added to the polymer to modify specific characteristics. Tints, dyes and pigments are added to change color or transparency. Antioxidants reduce yellowing in the process. Carbon black may be added for color and to absorb harmful ultraviolet light. Blowing agents may be added to create foam. Fillers are added to increase bulk and reduce costs. Some fillers such as calcium carbonate and clay also increase stiffness. Reinforcement such as glass and carbon fibers are typically added to increase physical properties such as tensile strength

Figure 5.23: Deep Draw Cup

and stiffness. It is critical to understand the impact of each of these materials on the formability of the polymer.

5.5   PART GEOMETRY

The specific part geometry should meet thermoforming design guidelines to create the best quality part. The following are some of these guidelines:

5.5.1   CORNERS AND RADII

A corner is the intersection of two or more planes. There are several types of corners in any part geometry. These include two-dimensional (2D) inside and outside corners and three-dimensional (3D) inside and outside corners. The material must be drawn into an inside corner and wrap over an outside corner. The part designer often wants sharp corners while the thermoform guidelines call for the largest radius possible.

Corners are also stress concentrators. As noted earlier, the last area of the sheet to touch the mold is the thinnest section. In a cavity mold this is the inside corner and the 3D corner is typically the last point to draw. As a result, the sharper the corner, the thinner and weaker the part will be at that point. If the corner can have a large radius or chamfer, the part will be thicker, stronger and have less stress. One source indicated that a minimum radius of 0.19" (4.8 mm) is acceptable while another suggested the following formula: r is equal to or greater than 3 times the thickness / 4. In this example a sheet thickness of 1/8" or 0.125" would have a minimum radius of 0.125" X 3 / 4 or 0.094". Yet another source uses a simple formula of 1.5 times the starting sheet thickness. The best rule in determining corner radii is to make it as large a possible and realize this will be a weak and high stress point in the part.

5.5.2   DRAFT ANGLES

A draft angle can be defined as the angle the mold wall makes with vertical. If the mold wall is vertical, there is 0 draft. Draft angels create the same design issues with thermoformed parts as they do with other plastics parts produced using other processes such as injection molding and compression molding. They may be very undesirable for the part designer but necessary for the process. The importance of draft angles is quite different for the cavity versus the plug style mold. As the heated, formed, sheet cools, it shrinks thus grabbing the plug but shrinking away from the cavity. Therefore, cavity molds do not necessarily require any draft although up to 2 degrees is suggested. Plug molds must have draft in order to allow the part to be released. Higher shrink materials, particularly crystalline materials such as polyethylene, may require more than 5 degrees while most amorphous materials such as polystyrene and polycarbonate may only require 3 degrees. If the mold is textured, the amount of draft for both plug and cavity molds increases depending on the type and depth of the texture. In addition, high forming pressures such as those used in pressure forming may also require greater draft.

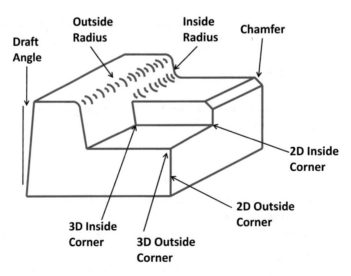

Figure 5.24: Example Part

It should be noted that many part designs actually have plug and cavity sections in the same mold. For example, a simple tray with dividers (Fig. 5.25) will have both plug and cavity areas to deal with. The best rule of thumb is to have as much draft as possible. Designers often incorporate the need for draft with the opportunity for stacking parts. Large draft angles also allow parts such as drinking cups to be stacked together.

5.5.3 DEPTH OF DRAW

Part geometry must consider the depth of draw required in the process to produce the part. As noted previously in this chapter, the decision whether to use a plug or cavity mold has a huge impact on draw depth as well as the location of wall thickness variation. Each individual part design needs to be evaluated for its depth of draw requirements. The simple tray with dividers is a good example. Figure 5.25 illustrates this type of divided tray. Each tray section is 1" deep.

Figure 5.25: Divided Tray Part Example

Figure 5.26:   Figure 5.27:

Regardless of whether this is fundamentally a plug or cavity mold, both options will contain plug and cavity sections. However, the draw ratio varies greatly between these two options. If the mold is designed with the dividers upright as seen in Fig. 5.26, the heated sheet will first touch the dividers, and then draw down into the cavity sections. In this case, the sheet must stretch from an unformed length of 1.75 to a formed length of 3.75" creating a linear draw ratio of 2.14. The bottom corners of the tray would the thinnest section.

If the mold was designed with the tray dividers down, as seen in Fig. 5.27, the heated sheet would first touch the tray bottom and draw into the dividers. In this case, the sheet must stretch from an unformed length of .25" to a formed length of 2.25" creating a linear draw ratio of 9 which exceed the draw ratios of all materials. In this case, the tray bottom would be as thick as the original unformed sheet and the dividers would be extremely thin, probably unusable.

5.5.4   WEBBING

Webbing is a phenomenon that occurs when the heated material makes contact with itself while forming and permanently bonds together causing a wrinkle. This is a significant problem at outside corners on plug molds when the part requires a deep draw. Webbing may also occur between plug areas within the same mold if these areas are too close together or are too tall. The part design should be reviewed for areas that are likely to create a webbing problem.

5.5.5   UNDERCUTS OR NEGATIVE DRAFT

Undercuts are commonly designed into thermoformed parts. For example, the polystyrene snap lid on a disposable coffee cup has an undercut to hold it onto the lip of the cup. Although this is a simple example, this snap fit must be tight enough to create a water tight seal yet loose enough to easily snap in place without damaging the lip of the cup or tearing the lid. In this case, no special tooling techniques are required to create this undercut. Since these parts are typically formed in a cavity mold, part shrinkage and an air blast are all that is required to remove the part. It is suggested that the undercut section not be greater than 0.030" below the mold surface for normal part removal. Of course this will vary with the rigidity of the part which is determined by the polymer in use as well as the part thickness.

Deep undercuts such as those shown in Fig. 5.28 may require more sophisticated tooling to remove the part. The tool may incorporate a simple hinged area that moves as the part is ejected from the mold Fig. 5.29 or more sophisticated core pulls actuated with hydraulics, pneumatics or electrical solenoids as shown in Figs. 5.30 & 5.31.

5.6   PART APPLICATION ISSUES

There are many application issues to consider when designing a part. The following are a few that tie directly to the thermoforming process:

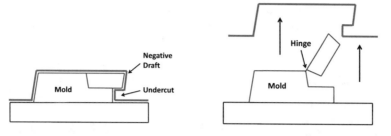

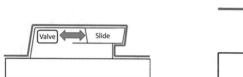

Figure 5.28: Undercut and Negative Draft

Figure 5.29: Hinged Release

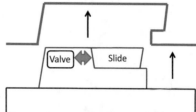

Figure 5.30: Slide System

Figure 5.31: Slide Release

5.6.1   USEFUL TEMPERATURE RANGE

Since thermoformed parts are formed rather than molded, the molecules are under stress. In fact, if a thermoformed part is placed in an oven and heated to its lower forming temperature, it may return to a flat sheet. This is called plastic memory. On the other hand, if this activity is performed on an injection molded part, it may twist and warp due to molding stresses but since the molecules flowed into their molded position, these forces are minimal. Therefore, it is important to understand the useful temperature range for each material to be selected for a thermoformed part and that this temperature is well below the minimum forming temperature.

5.6.2   STRENGTH REQUIREMENTS

The strength of any part is one of the fundamentals that any designer must consider. The strength may be measured in the amount of impact a part can withstand, the tensile strength of a material, the stiffness of the part, etc. The part geometry as well as the polymer selected are the primary determinants of the part strength. One of these measures of strength is discussed as follows:

5.6.3   STIFFNESS

The stiffness of a part is measured by its resistance to bending. The amount of stiffness required depends on the application of the part in its final use. In some applications, part stiffness is not an issue. For example, a thermoformed acrylic skylight could be very rigid or flex somewhat as long as it has enough rigidity and strength to carry the load of a heavy snow and withstand the impact of a hail storm. On the other hand, a disposable drinking cup must be rigid enough to be held while fully loaded yet this must be accomplished while maintaining the thinnest part possible to keep costs down. The stiffness may be increase using part geometry rather than increasing part thickness. Lips, corrugation, steps, domes, etc. are all geometric options to increase stiffness. (Fig. 5.32) Most thermoformed disposables and packaging incorporate these design options.

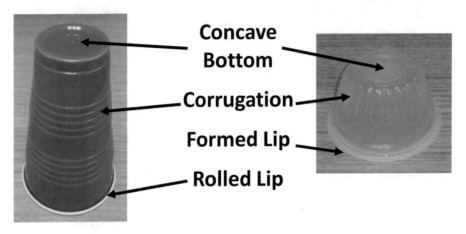

Concave Bottom
Corrugation
Formed Lip
Rolled Lip

Figure 5.32: Stiffening Techniques

5.7   QUALITY REQUIREMENTS

Overall quality control will be discussed in a later chapter, however quality requirements must be considered during part design. The following represents a few of these design issues:

5.7.1   COSMETICS

Are there specific cosmetic quality requirements? For example, the thermoformed housing for a point of purchase terminal may have very high cosmetic requirements on the outer surface as it is seen by the user and the customer. The base of this unit as well as the inside surface would have very minimal requirements. Smooth, high gloss surfaces are often the most difficult to create as any chill marks, discoloration, and other process issues are quite visible. Additionally, damage can occur during part removal, part trimming, and handling processes. Whenever possible a flat finish with a texture should be considered as this tends to hide some of these problems.

In addition to cosmetic requirements, there may be specific requirements for optical clarity.

5.7.2   OPTICS

Optical quality parts have the same issues as discussed above plus the light transmission may also be of concern. As thermoforming is a stretching process leaving the parts with internal stress and varying wall thickness, light transmission may be distorted. Chill marks and stress lines may be visible after forming or become visible after the part has been subjected to the environment over time.

5.7.3 DIMENSIONAL TOLERANCES

The designer must have some expectations for dimensional tolerances for thermoformed parts. Unfortunately the numbers of variables in this process make predictability somewhat difficult. Until a mold is made and materials tested using the process selected, actual process capability is difficult to determine. The good news is that inexpensive prototype molds can be made quickly to test the design and determine actual tolerances to be held. More about these prototype molds will be discussed in the next chapter on tool design.

Material shrinkage, process control and part geometry all play a role in the final part dimension. Different materials shrink at different rates and a specific material may also shrink at different rates based on internal stresses produced during the sheet manufacturing process and the amount of heat applied during the process. Sheet quality will be further discussed in the quality control chapter. During the process if more heat is applied, the sheet will expand more and therefore shrink more after forming. The amount of shrink may be reduced by lowering the mold temperature or using cooling fixtures to hold the part after forming.

A general rule of thumb for the designer to estimate tolerances for thermoformed parts has been determined as follows: Packaging and disposables made with thin gage materials should hold a process tolerance of plus or minus 0.010" for up to a 6" part. 0.001" is added to this tolerance for every inch above 6". Therefore, a 23" long part would have a tolerance of plus or minus 0.017". Thick gage non-packaging materials should hold a tolerance of 0.030" up to 12". 0.002 is added for every inch above 12". Therefore, the 23" part mentioned above would have a tolerance of 0.052".

CHAPTER 6

Mold / Tool Design

6.1 PURPOSE OF THE MOLD OR TOOL

The thermoforming mold or tool has many purposes. First and foremost it defines the geometry of the part being formed. In addition it may provide part texture, lettering, logos and other molded in details. The mold also acts as a heat exchanger to draw heat from the formed part to reduce cooling time. It may also provide locations for part trimming or even perform the trimming operation during the forming process know as in-mold trimming. The mold must be able to handle the forces repeatedly applied during the forming process as well as be robust enough to produce the desired number of parts for which it was designed. It must have adequate air flow to evacuate the space between the heated sheet and the mold surface during the forming process.

6.2 MOLD MATERIALS

A thermoforming mold may be made of a variety of materials as long as they meet the tool purpose as discussed above. Low volume molds, sometimes called prototype molds, are typically made to produce a few parts and/or test a part or process design. The purpose of these molds is low cost and quick build time. Materials include wood (Figs. 6.1 & 6.2), plaster, syntactic foam (Fig. 6.3) and thermoset materials like phenolic, epoxy and composite materials including fiberglass reinforced polyester. The biggest negative with all of these materials is their thermal conductivity, or actually their lack of conductivity. All are considered thermal insulators and barely meet the requirement of a heat exchanger. For example, the thermal conductivity of epoxy is 0.131 and aluminum 6061 has a thermal conductivity of 96.8. Simply stated, aluminum conducts heat over 700 times better than epoxy. For this reason, aluminum is the most used material for thermoforming production molds (Fig. 6.4). These molds are typically built using CNC machining technology (Fig. 6.7). Table 6.1 shows the thermal conductivity properties of various mold materials.

6.3 MOLD GEOMETRY / SHRINKAGE

The mold geometry follows all of the same guidelines as were stated in Chapter 5, Part Design, regarding draft, corner radii, etc. The mold must also be made to allow for part shrinkage so the mold is typically larger than the part. There are many variables that affect part shrinkage as discussed previously. These include, but are not limited to, the polymer in use, the forming temperature, the mold temperature, the orientation of the molecules from the sheet production, etc. The best method to determine actual shrink rates is to test the process or use similar designed parts and molds as

Figure 6.1: Simple Wooden Mold *Courtesy of Helton, Inc.*

Figure 6.2: Large Wooden Mold *Courtesy Tooling Tech. Group*

Figure 6.3: Syntactic Foam Mold

Figure 6.4: Aluminum Hood Mold *Courtesy of Helton, Inc.*

Figure 6.5: Thermoformed Hood *Courtesy of Helton Inc.*

Figure 6.6: Hood Installed *Courtesy of Helton Inc.*

Figure 6.7: CNC Mold Build *Courtesy Tooling Tech. Group*

Figure 6.1: Simple Wooden Mold *Courtesy of Helton, Inc.*

Figure 6.2: Large Wooden Mold *Courtesy Tooling Tech. Group*

Figure 6.3: Syntactic Foam Mold

Figure 6.4: Aluminum Hood Mold *Courtesy of Helton, Inc.*

Figure 6.5: Thermoformed Hood *Courtesy of Helton Inc.*

Figure 6.6: Hood Installed *Courtesy of Helton Inc.*

Figure 6.7: CNC Mold Build *Courtesy Tooling Tech. Group*

CHAPTER 6

Mold / Tool Design

6.1 PURPOSE OF THE MOLD OR TOOL

The thermoforming mold or tool has many purposes. First and foremost it defines the geometry of the part being formed. In addition it may provide part texture, lettering, logos and other molded in details. The mold also acts as a heat exchanger to draw heat from the formed part to reduce cooling time. It may also provide locations for part trimming or even perform the trimming operation during the forming process know as in-mold trimming. The mold must be able to handle the forces repeatedly applied during the forming process as well as be robust enough to produce the desired number of parts for which it was designed. It must have adequate air flow to evacuate the space between the heated sheet and the mold surface during the forming process.

6.2 MOLD MATERIALS

A thermoforming mold may be made of a variety of materials as long as they meet the tool purpose as discussed above. Low volume molds, sometimes called prototype molds, are typically made to produce a few parts and/or test a part or process design. The purpose of these molds is low cost and quick build time. Materials include wood (Figs. 6.1 & 6.2), plaster, syntactic foam (Fig. 6.3) and thermoset materials like phenolic, epoxy and composite materials including fiberglass reinforced polyester. The biggest negative with all of these materials is their thermal conductivity, or actually their lack of conductivity. All are considered thermal insulators and barely meet the requirement of a heat exchanger. For example, the thermal conductivity of epoxy is 0.131 and aluminum 6061 has a thermal conductivity of 96.8. Simply stated, aluminum conducts heat over 700 times better than epoxy. For this reason, aluminum is the most used material for thermoforming production molds (Fig. 6.4). These molds are typically built using CNC machining technology (Fig. 6.7). Table 6.1 shows the thermal conductivity properties of various mold materials.

6.3 MOLD GEOMETRY / SHRINKAGE

The mold geometry follows all of the same guidelines as were stated in Chapter 5, Part Design, regarding draft, corner radii, etc. The mold must also be made to allow for part shrinkage so the mold is typically larger than the part. There are many variables that affect part shrinkage as discussed previously. These include, but are not limited to, the polymer in use, the forming temperature, the mold temperature, the orientation of the molecules from the sheet production, etc. The best method to determine actual shrink rates is to test the process or use similar designed parts and molds as examples. Table 6.2 represents a starting point for mold designers to increase the mold size to accommodate for shrinkage.

Table 6.1: Thermal Conductivity	
Material	Thermal Conductivity
Aluminum - 6061	96.8
Stell - P20	20.0
Stainless Stell - 316	9.4
Plaster	0.174
Epoxy	0.131
Maple	0.094
Syntactic Foam	0.070
Air (used as a reference)	0.016

Table 6.2: Material Shrinkage Allowance	
Plastic Material	Mold increase (inches per inch)
HDPE	0.017 - 0.022
PET	0.007
PP	0.017
PS	0.005
PVC	0.004

Example: The information in Table 6.2 would suggest that the mold maker build a mold cavity for a 5" long part to be made of PVC (0.004 X 5) 5.020" long. If the same part was to be made of HDPE, the mold cavity would measure (0.017 X 5) 5.085" long.

6.4   VENTING

All thermoform molds must have a means for the air to be evacuated between the heated sheet and the mold. These are called vents for pressure forming and vacuum holes for vacuum forming. They basically serve the same purpose and the fundamental design principles are the identical. The primary purpose of these vents or vacuum holes is to facilitate the heated sheet to form and make contact with the mold surface as fast as possible. Although the term hole is used, these may be slots, channels or any other opening which allows air to flow. The size of these openings must be small enough so that the heated sheet will not form into the opening yet large enough to quickly evacuate the air. Ideally there will be no mark on the part from these openings. This, however, is less important on packaging and disposables and close examination of these products often reveal visible marks. The actual size also varies with the material being formed as well as the thickness of the material contacting the opening. As the volume of air needing to be evacuated increases, so do the number or size of these openings. Crystalline materials such as polyethylene and polypropylene are quite soft and fluid-like at forming temperatures. This characteristic requires small openings in the range of 0.003"– 0.015" so no mark will be left on the part. Molds used to form thick gage ABS may have openings up to 0.040" without leaving a mark on the part.

The vent or vacuum holes are located at the last points to form and where any air may be trapped between the sheet and the mold. To increase the air flow out of the mold, vent and vacuum holes are typically back drilled. This is accomplished by enlarging the hole on the opposite side of the mold from the cavity. The larger hole is commonly two to three times the diameter of the vent and vacuum holes. Figs. 6.8 and 6.10 illustrate the back drilling concept. Figs. 6.11–6.13 show an example of back drilling on a mold insert.

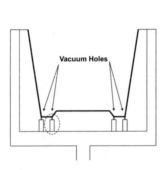

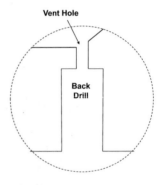

Figure 6.8: Mold with Vent Holes

Figure 6.9: Vent Hole Location

Figure 6.10: Back Drill

Actual Vent Size = 0.011"

Actual Back Drill Size = 160"

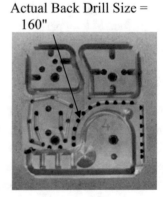

Figure 6.11: Mold Cavity Insert

Figure 6.12: Vent Hole Size

Figure 6.13: Back Drill Size

6.5   TEMPERATURE CONTROL

Most aluminum molds have some type of control to maintain the correct temperature for the material being formed. The most common temperature control method uses water flowing through a controller that heats or cools the water to maintain a preset temperature. Water channels are machined into mold to most effectively remove heat. Packaging and disposables may have molds cooled just above the temperature at which condensation forms on the mold while more exotic materials like polysulfone require mold surface temperatures reaching 300 degrees F. Fig. 6.14 illustrates cooling channels machined through a mold cavity. The path of the coolant should be designed such that equal cooling occurs within the mold. Fig. 6.15 illustrates a manifold system that would more evenly cool the mold. Fig. 6.16 illustrates a poor design where the fluid increases in temperature as it flows through the mold creating uneven cooling.

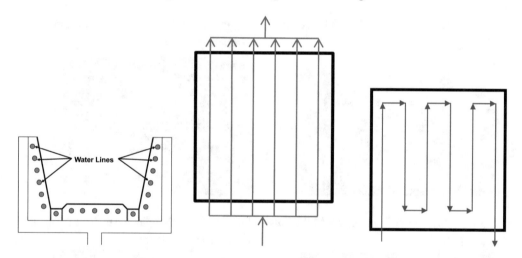

Figure 6.14: Mold Cooling

Figure 6.15: Cooling Flow - Good Example

Figure 6.16: Cooling Flow - Bad Example

6.6   CAVITIES

Basic thermoforming molds include a cavity or plug often machined into a block of aluminum as shown in Fig. 6.17. More sophisticated molds have standard mold bases with cavity inserts added, such as the insert shown in Fig. 6.11. High volume parts are rarely produced one at a time. Rather they are produced in multi-cavity molds as shown in Fig. 6.19 & 6.21.

These may be used in sheet feed or roll feed machines. Another cavity option is known as a family mold which consists of several different cavities producing parts which act as a set of parts in the end product. For example, a computer housing top and bottom may be made in at the same time in a family mold. The assumption is that no tops would be needed without also needing a bottom.

Figure 6.17: Large Single Plug Mold *Courtesy of Helton, Inc.*

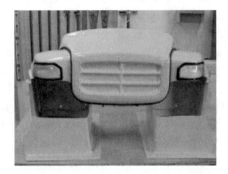

Figure 6.18: Sample Part From Fig. 6.17 *Courtesy of Helton, Inc.*

Figure 6.19: Thirty Two Cavity Mold *Courtesy Arvind Polymers*

Figure 6.20: Sample Parts From Fig. 6.19 *Courtesy Arvind Polymers*

Figure 6.21: Multi-Cavity Mold *Courtesy Brown Machinery LLC.*

Figure 6.22: Large Two Cavity Family Mold - Left & Right Fenders *Courtesy of Helton, Inc.*

Fig. 6.22 shows a type of family mold for off road vehicle fenders. The mold creates a left and right side fender at the same time. Cavities may also be made of a different material than the standard mold base as a cost saving measure. The base may be a lower cost material as it does not require the same characteristics as the cavity. The mold may also have replaceable inserts in high wear areas that may need to be replaced without the need to build an entire mold. One example of an insert may be a trim blade for in-mold trimming. As the blade wears and gets dull, it is simply replaced.

CHAPTER 7

Quality Control Issues

7.1 INTRODUCTION

The control of quality is key to the success of any business. Thermoforming is no exception. Since thermoforming is a secondary process requiring sheet material to be produced first, one critical area of concern is the quality of the incoming materials. As the old saying goes: "You cannot make a silk purse out of a sow's ear." Supplier quality will be discussed in the following categories: material source, material variation, orientation, moisture and cosmetics.

7.2 MATERIAL SOURCE

The supplier has many options regarding the source of the material to be used to create the sheet to be purchased. It is important to agree on these options with the supplier to provide greater consistency in the sheet material. Thermoforming could not exist as an economically viable process if the trim and scrap were not recovered and re-used. As material costs are typically the largest portion of the cost of a thermoformed product, it is important to keep material cost as low as possible without negatively affecting the product quality. Blending the recovered trim and scrap with new material is one such way to reduce material costs. The recovered material is called regrind and the new material is called virgin as it has not yet been formed. The sheet manufacturer has many variables to consider when blending regrind with virgin and many of these variables directly affect the thermoforming organization. Some of these issues are as follows:

7.2.1 REGRIND

Regrind quality control is always a big concern. Was the material kept clean and isolated to eliminate contamination of other materials including dirt, water, oil, other regrind, etc.? If not, these contaminates will become part of the sheet material and eventually the thermoformed part.

As thermoplastics are heated, stretched and reground, there is some deterioration in the physical properties. The history of the regrind being used is often unknown to the sheet manufacturer as well as the thermoformer. If material has been over exposed to heat it is weaker and may be also be "yellowed."

The use of regrind and the percent allowable needs to agreed upon between the supplier and thermoformer. Some products, such as medical, biological, food containers, etc. may limit the use of regrind and may not allow any at all.

7.2.2   SHEET ORIENTATION

As the sheet is being produced, typically using the extrusion process, molecules tend to become aligned or oriented. (Fig. 7.1) To reduce costs, the extruder, used to produce the sheet, is run as fast

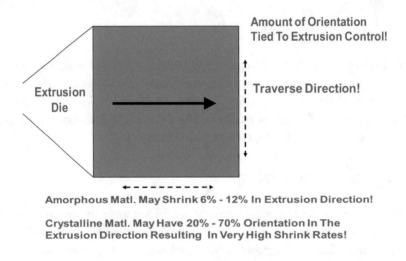

Amount of Orientation Tied To Extrusion Control!

Extrusion Die

Traverse Direction!

Amorphous Matl. May Shrink 6% - 12% In Extrusion Direction!

Crystalline Matl. May Have 20% - 70% Orientation In The Extrusion Direction Resulting In Very High Shrink Rates!

Figure 7.1: Sheet Orientation

as possible. Unfortunately this is one of the causes of orientation in the sheet which is undesirable to the thermoformer. The goal should be 0 – 5% orientation in most cases. If molecular orientation exists in excess, the sheet will have internal stress that will affect both the thermoforming process as well as the thermoformed product. When heated the stress may cause shrinkage and the sheet may actually pull out of the clamping frame. The orientation of the sheet will also result in excessive orientation on the product which may cause warping, twisting and stress cracking.

7.2.3   ORIENTATION TEST

Incoming sheet can quickly and easily be tested for orientation and stress by performing the following test which is indisputable:

Cut a section of the sheet to a 12" square. This may be done several times in different areas of the incoming sheet including the edges and the center. Using a marker, draw parallel lines in one direction across the entire sample about $\frac{1}{2}$" apart. (Fig. 7.2)

Cut the sheet into 6 samples as shown in Fig. 7.3. Each sample is 2" X 7". Measure the samples in width and length. Preheat an oven to the normal forming temperature for this material. Place the samples on a non-stick sheet or screen and place it in the oven for the normal pre-heat time. (Fig. 7.4)

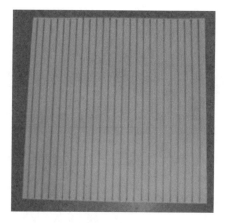

Figure 7.2: Test Sheet

Figure 7.3: Test Samples

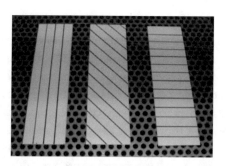

Figure 7.4: Samples Placed in Owen

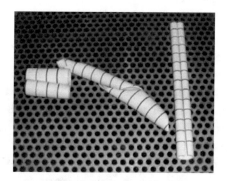

Figure 7.5: Highly Oriented Sample - 0.060"
Polystyrene

Remove the sheet and let the samples cool. Observe the samples. If high orientation stress is present the parts will curl tightly. If moderate orientation exists the samples will only slightly curl. If the samples do not curl, they can be measured and compared to the pre-heat measurements.

Fig. 7.5 shows a sample of 0.060" polystyrene curled significantly when heated to normal forming temperature of 300 degrees F. This indicates an unacceptable level of sheet orientation produced during the extrusion process when the sheet was produced. This lot of sheet stock will be returned to the supplier.

The samples shown in Fig. 7.6 of 0.030" polystyrene were cut from the same sheet. This material is used for packaging. The sample on the left was heated to normal forming temperature, 300 degree F. The sample on the right was heated to the upper forming temperature of 360 degrees F. Greater amount of stress was indicated as the temperature was increased. Fig. 7.7, a sample of 0.100"

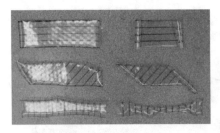

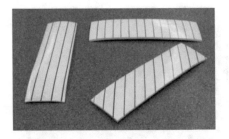

Figure 7.6: Polystyrene 0.030" - Parts on Left - Normal Forming Temp. - Parts on Right - Upper Forming Temp

Figure 7.7: ABS 0.100" Shows Little Orientation

ABS, showed little sign of orientation or stress as it was heated to its upper forming temperature of 360 degrees F.

7.2.4   MOISTURE

If moisture, even at the molecular level, is present in the sheet, it will form steam or vapor when heated. Some materials are hygroscopic and tend to absorb moisture at the molecular level. Examples of hygroscopic polymers include acrylic, polycarbonate, nylon, polyethylene terephthalate (PET), and acrylonitrile butadiene styrene (ABS). The sheet manufacturer must use selected drying procedures to prepare this material for extrusion or the sheet will have a variety of defects. If the sheet material is not stored properly or is stored for extended periods of time, it will also absorb moisture. The best practice is to tightly seal these polymer sheets in a moisture barrier such as polyethylene film at the extrusion process. If the sheet must be dried before the thermoforming process, added costs are incurred. Table 7.1 shows typical drying times and temperatures of sheet materials. The drying times will increase as sheet thickness increases.

Table 7.1: Sheet Drying Temperatures			
Polymer	Typical	Drying Temperature	Typical Drying Time*
	F	C	hrs
APET	150	65	3-4
CPET	320	160	4
ABS	175	80	3
Acrylic	175	80	3
PC	300	150	4
*Drying times will vary with sheet thickness			

7.2.5 COSMETICS

If a sheet has cosmetic defects before forming, these may very well be exaggerated after stretching. These defects may be the result of sheet manufacturing such as moisture bubbles and haze, contamination from dirt, oil, etc. Other defects such as scratches may also be from handling at the supplier, during shipping or in your own facility. Cosmetic inspection should be including in the receiving process to isolate the ownership of these problems. Procedures for proper handling and storage must also be part of the operating principles of the organization.

7.3 PROCESS QUALITY

The quality of the manufacturing process must be controlled to achieve consistency in the quality of the finished product. There are many variables within the thermoforming process that affect quality. Some of these are discussed as follows:

7.3.1 MACHINE SET-UP

Machine set-ups should be documented and the documentation should be consistently followed. As changes are made in the process to accommodate the variation in material, humidity, etc. these should also be documented and saved. All relevant process parameters should be specified including heater controls, actual sheet temperature, water temperature and flow, actual mold temperature, plug temperature, vacuum and or pressure, etc. It is wise to invest in several non-contact laser pyrometers to easily check actual temperatures. (Figs. 7.8 & 7.9) Of course training is needed for all set-up personnel and operators to reduce operator variation.

Figure 7.8: Non-Contact Laser Pyrometer

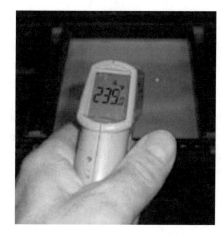

Figure 7.9: Pyrometer Reading

7.3.2   VARIATION

The quality purist will make the argument that no two things are alike and if we can measure accurately enough, the differences can be found. However, the practical application to variation is to accept that it exists and then do what we can to reduce it to an acceptable level for our application. As noted many times in the book, thermoforming is a stretching process and the products will have wall thickness variation dependent on the amount of stretching that occurs. Therefore, we will always have "in piece variation." "Part-to-part variation" is a term for the variation that occurs between parts. There are many causes of this variation including process, material, operator, environment, etc. Each of these causes needs to be monitored and controlled to reduce piece-to-piece variation. "Time-to-time variation" is that variation that occurs between work orders, set-ups, production runs, etc. Is there a difference in the products made last month that those made today? Time tends to make variation more visible. There is a new set-up technician, a change in the tool, a new supplier or a different batch of material, different amounts of regrind, changes in temperature or humidity due to seasonal changes, etc. Time-to-time variation is often much harder to control but must be monitored.

7.3.3   TOOL QUALITY

Careful handling and storage of molds can eliminate a lot of headaches in production. Each mold should have a specified storage location which protects it from being damaged in any way. It is a good practice to save the last part produced and thoroughly inspect the part and the mold before storing the mold. Make any repairs at this time so all stored molds are ready to set-up when needed. These issues also apply to all secondary tooling such as punch dies and trim fixtures.

7.3.4   FACILITY QUALITY

The quality of the product can certainly be affected by the quality of the facility. Cleanliness and organization can have a direct impact on quality and profitability. If grinding is performed in the facility, keep is out of the production and sheet storage areas if possible. The dust and fine strands, also known as fines, can contaminate the process causing quality problems. Keep the regrind clean and dry as not to induce problems for the sheet supplier. Keep water and oil off of the floor and equipment as these are quality as well as safety issues. Know and follow all local, state and federal safety regulations. Many of these directly affect product quality as well. More ideas on facility quality is discussed in Chapter 8, Lean Operations.

7.4   QUALITY INSPECTION

Ideally there is no need for inspection if the material and processes are well controlled. Reality as well as contractual obligations may create a different situation. Part inspection is often focused on the process start-up after a set-up is complete. Parts are monitored until the expectations are met consistently. Problem parts may be more regularly monitored based on historical issues of non-conformity. It is recommended that an inspection procedure be developed and documented and that this procedure be followed. Part inspection may be destructive such as cutting a part in half to measure specific dimensions or breaking the part during a falling dart impact test. More typically, non-destructive inspection is implemented. These inspections may include checking part thickness in critical areas, visible inspection of surface quality, color, cracking, etc. Go/ No-Go fixtures can be designed for very rapid inspection of overall dimensions and openings after secondary operations are complete. For more sophisticated parts, a coordinate measuring machine (CMM) may be used for complete dimensional analysis.

7.4.1 QUALITY TOOLS

It is beyond the scope of this text to discuss in detail the many quality assurance tools available to the processor. However, several of these tools will be presented in a general manner in the following section. More specifics are readily available on-line, in quality textbooks, and in a variety of quality training manuals.

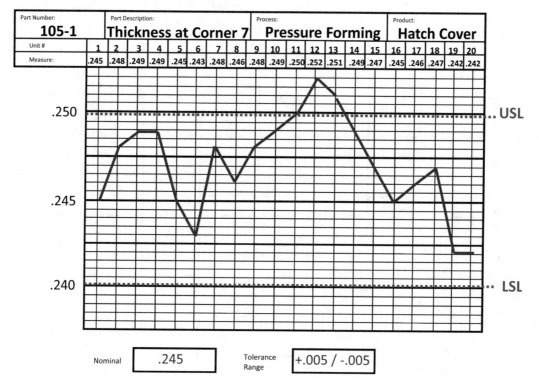

Figure 7.10: Run Chart Example

Quality characteristics can be divided into two categories: Variable data and Attribute data. Both types of data are readily available in any manufacturing situation. Variable data are those that can be measured such as distance, weight, hardness, etc. There are many quality tools that can be used to organize and analyze this type of data including Run Charts which simply record and graph the data in order of production looking for trends and out of specification situations. Fig. 7.10 shows a simple example of a Run Chart for part thickness in a specific area. There are also a variety of statistical process control tools including the Xbar & R Chart which determines and charts the sample means as well as sample ranges for analysis. Attribute data are those that cannot be measured but are simply good or bad, pass or fail, are present or absent, etc. The use of a go/no go gage

| PART 106-1 | | TIME | 1-3 PM | | | OPERATOR | J.D. |

Part #	Chill Mark	Tear	Insufficient Draw	Blister Bubble	Scorch Mark	Webbing	Other
1	X		X				
2	X		X				
3						X	
4							
5		X		X	X		
6							
7		X		X	X		
8							
9							DIRT
10	X						
11							
12	X		X				
13	X						
14							CHIP
15							
16	X		X				
17	X						
18							
19							
20	X						

Figure 7.11: CHEKSHEET

would be an example of a attribute. Although the part could be measured, it is determine that a simple pass/fail decision meets the needs of the particular application. Using a P-Chart (percent or proportion), the part is documented as either passing or failing. Often judgment is required on the inspector's part to determine pass or fail conditions. For example, cosmetic inspection may require the inspector to determine whether a part is in conformance or out of conformance based on their judgment or opinion. The data collected on a Checksheet is often presented on a Pareto Chart or Pareto Diagram. This quality tool presents the data in descending order from most frequent to least frequent for the purpose of setting priorities. This tool is based on the 80/20 rule which states that 80% of the effect is from 20% of the causes. Therefore, if we can eliminate the top 20% of the problems, 80% of the effect can be eliminated. Fig. 7.12 is an example of a Pareto Chart produced form the data collected on the Checksheet. (Fig. 7.11) In this example, if the cause of Chill Marks can be identified and eliminated, eight defective parts would be eliminated. On the other hand, if we used our valuable and limited resources identifying and correcting the cause of Scorch Marks, only two defective products would be eliminated. Pareto Charts can be an excellent tool for managers to communicate problems in a concise, summary format as well as monitor the results of corrective actions.

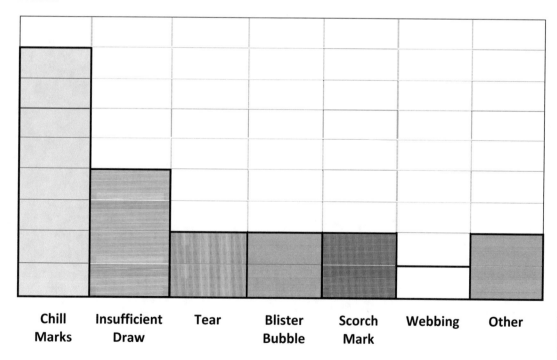

Figure 7.12: Pareto Chart

CHAPTER 8

Lean Operations

8.1 INTRODUCTION

Many organizations have adopted Lean operating techniques to reduce cost, improve quality and improve delivery performance. A fundamental question to ask is why are we doing things that do not add value to the product? If these activities do not add value, they do add cost! Also Lean improvements are everyone's responsibility not just management or engineering. Train everyone on these concepts and expect everyone to participate. In this chapter, several Lean concepts will be discussed and will be applied to the thermoforming operation. These include: Eliminating Waste, 6S, Process Flow Charts, Process Mapping, Visual Controls, Set-Up Reduction and Poka Yoke.

8.2 ELIMINATING WASTE

A fundamental concept in Lean thinking is to identify and eliminate waste. The following are examples of waste commonly found in a manufacturing environment:

1. Inventory – cost, space, security, obsolescence, shrinkage, etc.

2. Over Production – producing more parts than actually needed "just in case."

   Produce what is actually needed!

3. Over Processing – Spend only what time is needed to accomplish the task. For example: if a sheet requires 30 seconds to reach forming temperature, don't use 40 seconds. Understand the times required to accomplish tasks and compare the actual time to the expected time and adjust.

4. Unnecessary Motion – Remember that motion = time = money. Identify opportunities to reduce time wasted walking, searching, reaching, sorting, etc. Always think about ways to reduce motion.

5. Waiting – Wasted motion does not add value and neither does idle time. If someone or something (such as a process step) is idle, investigate the cause. This is not to imply that people must not stop working as we may expect is the practice in a sweat shop in a third world country. This concept is simply to identify opportunities to reduce time wasted waiting.

6. Rework / Correction – Any time and energy spent fixing something that was not done correctly is waste. These activities add cost rather than value to the product.

7. Continuously improve the processes. For example, go to trade shows and find routers that cut faster and leave a better edge than what you currently use. Continue to learn!

8.3   6S

As organizations move toward lean thinking, many adopt what is known as the 6S approach. This is an excellent philosophy to apply in manufacturing facilities to improve and maintain quality and reduce waste. The 6S techniques are as follows:

1. **Sort**: Eliminate items that are not needed. This may include molds, fixtures, tools, packaging, materials, etc.

2. **Straighten**: Everything should have its place and all should be labeled. Arrange things by their frequency of use. Keep things close that are used most frequently. Also keep things visible so nobody spends time hunting.

3. **Shine**: Keep everything clean. Set high standards. Make this a daily routine.

4. **Standardize**: Monitor the first three techniques. Standardize processes and equipment whenever possible.

5. **Safety**: Identify and eliminate dangerous or hazardous conditions in the work place.

6. **Sustain**: Make the above five techniques a way of life for everyone in the facility.

8.4   PROCESS FLOW CHART

Process Flow Charting an operation is an excellent method to identify what is really happening rather that what we think is happening. This starts with selecting a process to evaluate which may be machine set-up, material handling, trimming, or even a paperwork process such as work order generation and tracking. It is wise to use individuals that actually do the work in the area to perform the evaluation. Sometimes teams are used composed of people from different shifts or each shift may do their own evaluation which often explains the differences between shift outputs.

Fig. 8.1 is a simple example of a flow charting tool. It is clear that we want to use this tool to identify waste and other non-value added activities. Typically, tasks are documented in logical order and distance and time required for each task is noted. Following the documentation of the actual process, the data is analyzed for opportunities to reduce non-value added activities such as waiting or movement.

8.5   PROCESS MAPPING

Process Mapping is another tool used to identify waste, particularly focusing on flow and distance. A simple floor plan is drawn to scale. This can easily be done on graph paper or using computer

Part Number 105-1
Part Name - Hatch Cover
Process - Machine Set-up

Date - 2/14/XX
Operator - J.D.

DESCRIPTION OF OPERATION	Tooling/Equip	Distance/Ft	Time/Hrs
Transfer mold from tool crib	Lift Truck	300	0.15
Hang mold in machine XYZ	Hoist		0.2
Connect Coolant Lines			0.15
Heat Mold to 150 F.	Temp. Controller		0.25
Transfer skid of material from warehouse	Lift Truck	800	0.3
Set clamping frame per documentation	Air Wrench		0.1
Set oven temperatures - pre-heat to 325 F.			0.1
Set mold height	3/4" Wrench		0.1
Set process timers			0.1
Wait for ovens to reach 325 F.			0.25
Produce test part			0.1
QA test part			0.1
If acceptable, release machine to production			0.1
		Total Time	2.0 hrs

Figure 8.1: Process Flow Chart

programs such as VISIO, AUTOCAD, etc. The activities identified on the Process Flow Chart are drawn on this map to show the distance traveled to complete the task. Often the distance required is quite amazing and easy to reduce. Be sure the operators are involved in the data collection as well as the improvement activity.

8.6 VISUAL CONTROLS

The use of using visual controls is another tool to identify and eliminate waste. Wasted time and motion searching for a tool, for example, can easily be eliminated by having the needed tools at the point of use. Shadow boards are an excellent way to visually identify the tools location and availability. Piling tools in a tool box or drawer requires the operator to search and find what they need. A tool panel allows the operator to quickly grab what is needed and return it when finished.

The use of tool panels also reduces the loss of tools as missing tools are quite visible. Other visual controls include clearly labeled inventory storage locations, mold storage location, material to be reworked, etc. Figs. 8.2 & 8.3 show an example of a clearly labeled tool panel. Other examples of Visual Controls include clearly labeled locations for scrap, rework, queues, KanBans, etc.

Figure 8.2:   Figure 8.3:

8.7   SET-UP REDUCTION

Machine set-up is a very costly operation. It is non-value added time, requires high skilled people, and reduces capacity of the operation. There are many set-up reduction methods available including "single minute exchange of dies" (SMED) which has been used for many years. The following are some general concepts that can be quickly applied in a thermoforming operation to reduce this non-value added activity.

Separate the internal activities from the external activities. Internal activities must be done with the machine down such as setting the mold in the machine. External activities can be done before the machine is shut down for set-up. These include bringing the mold to the machine, connecting required fittings, hanger bolts, etc. The set-up procedure should be modified to have all external activities performed before the machine is shut down.

Standardize everything possible. Use only one size bolt if possible. Use identical water fittings an all molds and machines and be sure there are enough of them so they stay on the molds. Make as many attachment points and clamps quick and easy to change. Locate standard tools at each machine.

Train all set-up personnel in standard operating procedures (which they helped develop). Document the procedure for set-up noting differences in machines and molds. Monitor the set-ups and compare to the standard procedure. Adjust the procedure as needed to keep it accurate most efficient.

8.8   POKA YOKE

The term poka yoke simply means mistake proof. Investing the time and energy to develop processes that cannot be done wrong has huge payoffs in the future. There are many methods and techniques for mistake proofing and each tie to the specific application. A good starting point is to analyze scrap and rework. Is the scrap or rework a result of something being done wrong the first time? If so, this may be a good candidate for Poka Yoke. For example, if a trim fixture is placed on a part incorrectly, the routing process will ruin the part. A Poka Yoke example would be to design the fixture so it can only fit one way, the correct way. Mistake proofing a process can be a great team activity and often requires a great deal of creativity. This can be a very rewarding activity for the operators to be involved.

CHAPTER 9

Environmental Issues

9.1 INTRODUCTION

Plastics have been the focus of many environmental discussions over the years. Some of the issues are justified and many are not. For example, disposable plastics packaging is often sited as an environmental issue due to the fact that it does not break down and it is filling the landfills to capacity. However, the fact that this packaging extends to storage life of food products creating less waste is rarely discussed. Also plastic packaging takes up about 8% of the landfill space and much of this waste can be recycled if the consumer chooses. The intent of this chapter is not to defend the plastics industry but is to discuss issues that may affect the thermoforming sector. These include source reduction, recycling, and material selection.

9.2 SOURCE REDUCTION

Source reduction refers to the reduction of material used, therefore reducing the amount of scrap or trash produced. The use of plastics for packaging substantially reduces the amount of trash when compared to other alternatives such a paper, cardboard and metal containers. Additionally, innovations in plastics materials and processes continually reduce the volume of plastics used for a particular application. Today's packaging is thinner, yet as strong as packaging a decade ago. Although the driving force behind these changes is typically economics, the environment is also a winner. Additionally, the less packaging weighs, the less fuel is consumed and less emissions are created during transportation.

9.3 RECYCLING

Recycling can be divided in two sectors: in plant and post consumer recycling. As noted previously in this book, the thermoforming industry could not exist as an economically viable enterprise if the scrap and trim were not recycled and reused. Since its inception, the thermoforming industry has been a leader in the recycling effort because it simply made good business sense. Today's thermoforming operation discards little plastic scrap and even more can be saved from the landfill with improved housekeeping and material handling procedures.

Post consumer recycling is an entirely different issue. Although many packaging materials and disposables are recyclable, the desire to do so must exist as well as the infrastructure to collect, sort and reclaim these items. Most communities have recycling programs that accept specific materials such as PET and HDPE. Many thermoformed packages and disposables are made from other materials such as polystyrene (PS), polystyrene foam, polypropylene (PP), and polyvinyl chloride

(PVC) which are often not collected for recycling because there is little demand for these materials in the recycling industry.

The plastics packaging industry created the recycling labeling process several decades ago to facilitate the consumer's ability to identify different types of plastics making the sorting process more convenient. These labels are formed directly into the thermoformed package as well as many other disposables. Table 9.1 shows this system of labeling and Fig. 9.1 shows the recycle symbol that is formed on all packaging and many disposable products.

Figure 9.1: Recycling Symbols

Table 9.1: Recycling Labeling System		
Plastic Type	Recycle Acronym	Recycle Number
Polyethylene Terephthalate	PET or PETE	1
High Density Polyethylene	HDPE	2
Polyvinyl Chloride	V	3
Low Density Polyethylene	LDPE	4
Polypropylene	PP	5
Polystyrene	PS	6
Other	OTHER	7

9.4   MATERIAL SELECTION

The choice of materials affects both source reduction and recycling. Stronger, thinner materials create less volume of scrap and particular materials are more likely to be recycled than others. Because of the high demand for recycled PET, the infrastructure to recycle this material is well in place in most communities. The highest volume of this material comes from soda bottles and other blow molded containers. The demand for post consumer recycled PET is due its use in non-food contact products such as polyester fiber for carpeting and clothing. Because the recycling system is in place for this material, many product designers have moved away from PVC and HDPE and selected PET when it can meet the product requirements. Multilayer sheet may also be used with membrane layers to enhance material properties to meet a particular requirement.

9.5   GREEN PLASTICS

The definition of "green" is ever evolving. One definition focuses on the "end of life" of a product. For example, will the product readily biodegrade? Another definition may focus on the environmental issues with processing the materials including the energy, water and chemicals required in manufacturing, pollution generated during processing, etc. One of the latest areas of focus, particularly for biopolymers, is the source of the carbon to produce the polymer rather than the end of life. In this case the manufacturing of plastics is from renewable sources rather than hydrocarbons such as petroleum, coal and natural gas.

Green plastics have been increasing in usage for several years in the packaging and disposable sector. These are typically materials that will break down in a reasonable amount of time. They are also often considered sustainable as many are made from biological sources such as corn and soy beans. Some are marketed a biodegradable, photodegradable and/or compostable. Photodegradable HDPE has been used to produce six-pack can holders for over a decade. If the six-pack holder contains a small diamond symbol, it is photodegradable. These materials are intended to break down into small particles in about three weeks when exposed to outdoor sunlight. The plastic industry, particularly the recycling sector, has expressed concern about these materials as they can contaminate the recycling stream, causing even more scrap. However, the use of these materials is on the rise and the industry must determine a system to properly handle these new materials.

The demand for green plastics will continue to rise as companies like Toyota and NEC have corporate goals to replace current materials with green plastics. By the year 2020, Toyota expects to have at least 20% of its plastics in use to be green. Much of this will come in the increased use of Polylactic Acid (PLA) for interior panels. PLA is a corn based plastic material and its use is growing exponentially. Other fast growing biopolymers include polyethylene and polyvinyl chloride produced from Brazilian ethanol which is created from sugar cane. It is reported that these materials are "indistinguishable" from petroleum derived resins. A green version of polypropylene is also available which contains 50% starch blended with 50% polypropylene.

Biopolymers are not a new material. I fact the first plastic material developed was a biopolymer. Cellulose nitrate, patented in 1869, was derived from plant fiber known as cellulose. Cellulosics grew into an entire family of resins including cellulose acetate best known as photographic film cellophane tape, cellulose acetate butyrate, best known as tough handle for tools such as wood chisels.

Cellophane is still in use today after reaching its peak usage in the 1960's. It can still be found in some candy wrappers, as a transparent moisture barrier on cigarette packaging and frames for glasses.

Figure 9.2: Cellulose Acetate

Figure 9.3: Cellophane Tape

Figure 9.4: Cellophane Wrapping

Figure 9.5: Cellophane Wrapping

Figure 9.6: Cellulose Frame

Figure 9.7: Henry Ford's 1941 Plastic Car

Beginning in the 1920's, Henry Ford experimented with using soybeans to produce a polymer material for automotive body panels. There was success in using this material for steering wheels, interior trim and dashboard panels; the program was dropped with the onset of World War II. He did produce a prototype in 1941. (Fig. 9.7) Following the war, there was an abundance of petroleum as well as the infrastructure to develop petrochemicals; therefore this program was never continued.

Bibliography

[1] Advanced Extrusion, *Standard Thermoforming Equipment Overview.* 16 Jan. 2008 `<http://www.advancedextrusion.com/docs/standard-thermoforming-equipment.htm>`.

[2] Alexander, Judd H. *In Defense of Garbage.* Westport. Ct: Praeger Publishers, 1993.

[3] Askin, R.G. and Goldberg, J.B. *Design and Analysis of Lean Production Systems*, New York: John Wiley & Sons, Inc., 2002.

[4] Formech International. *A Vacuum Forming Guide.* 18 Dec. 2008 `<http://www.formech.com>`.

[5] Harper, Charles A. (editor) *Modern Plastics Handbook.* New York: McGraw Hill, 2000.

[6] Illig, A. *Thermoforming – A Practical Guide*, Munich: Hanser Publishers, 2001.

[7] McConnell, W.K. *Thermoforming Technology Society of Plastics Engineers Seminar Book*, Fort Worth, TX: McConnell Co, Inc., 1994.

[8] NKS/Factory Magazine. *Poka-Yoke: Improving Product Quality by Preventing Defects.* Portland, OR: Productivity Press, 1988.

[9] Richardson, T.L. *Industrial Plastics: Theory and Application* (2nd ed.), Albany, NY: Delmar Publishers Inc., 1989.

[10] Rosen, S.R. *Thermoforming: Improving Process Performance*, Dearborn, MI: Society of Manufacturing Engineers, 2002.

[11] Schut, Jan H. "What's Ahead for Green Plastics?" *Plastics Technology* Feb. 2008. 64–71.

[12] Stevens, E.S. *Green Plastics: An Introduction to the New Science of Biodegradable Plastics.* Princeton, NJ: Princeton University Press, 2002.

[13] Stevens, E.S. "What Makes Green Plastics Green?" *Biocycle* Mar. 2003. 24–27.

[14] Strong, A.B. *Plastics Materials and Processes* (3rd ed.), Upper Saddle River, NJ: Pearson Prentice Hall, 2006.

[15] Throne, J.L. *Thermoforming*, Munich: Hanser Publishers, 1987.

[16] Throne, James L. *Thermoforming 101.* 23 Feb. 2009
      <http://www.thermoformingdivision.com/thermoforming101/index.htm>.

[17] Throne, J.L. *Understanding Thermoforming*, Munich: Hanser Publishers, 1999.

[18] Throne, J.L. *Understanding Thermoforming* (2nd ed.), Munich: Hanser Publishers, 2008.

[19] Wikipedia. *Recycling.* 4 Jan. 2009
      <http://en.wikipedia.org/wiki/Recycling>.

[20] Wikipedia. *Resin Identification Codes.* 4 Jan. 2009
      <http://en.wikipedia.org/wiki/Resin_identification_code>.

About the Author

PETER W. KLEIN

Dr. Peter W. Klein is Chair of the Department of Industrial Technology in the Russ College of Engineering & Technology at Ohio University in Athens, Ohio. He has been a faculty member since 1990 and has been responsible for the development and teaching of a variety of courses including: Industrial Plastics, Plastics Forming & Fabrication, Plastics Molding Processes, Quality Assurance, Introduction to Manufacturing, Introduction to Manufacturing Processes, and Product Manufacturing. He has also taught Operations Management for the College of Business at the undergraduate, graduate, executive MBA, and international MBA levels. Prior to joining the faculty at Ohio University he held various management positions in the manufacturing operations areas within Hewlett Packard, IBM, and AMPEX corporations. He began developing and teaching classes in plastics processing in 1975 at Colorado State University.

He has authored numerous publications in the areas of plastics processing, operations management, and quality assurance. He regularly makes presentations on these topics at national and international conferences. He is a member of the Society of Plastics Engineers, Association of Rotational Molders International, American Society of Engineering Educators, and the National Association of Industrial Technology.